Unleash your Engineering Systems Power

By: Dr. Federica Robinson-Bryant

Unleash your Engineering Systems Power Published by Denotion Research Group
www.DenotionResearch.com

ISBN:
978-1-958634-30-1 (Color Paperback)
978-1-958634-29-5 (B/W Paperback)
978-1-958634-22-6 (eBook)

Printed in United States 1st Edition

Dedicated to the powerful forces around me, especially those who truly aspire to witness the success of others.

Contents

Preface
Issuing a Resounding PSA

There is nothing more difficult to take in hand, more perilous to conduct, or more uncertain in its success, than to take the lead in the introduction of a new order of things.

- Niccolo Machiavelli

Key Topics

- What is the core goal of this book?

- How is the book structured to achieve its objective?

Greetings, Future Engineering Systems Enthusiast!

Congratulations on taking the first step towards a path of endless possibilities.

This book offers a range of insights that you may find both engaging and thought-provoking. The material may require some initial effort, as I am a "mature" lady attempting to pierce your young mind and ignite your curiosities. Yet, I challenge you to read and keep reading… somewhere along the way, you should let go of your inhibitions and realize this text is not about steering you in one direction or another but instead, sharing insights gained along my own journey, so that you are not only aware of the existence of the invisible bounds you currently exist within but also can unleash things within you and around you to really master your own journey.

Who am I, you ask?

Not so long ago, I was some version of you.

Today, I am a relatively older lady still blazing trails, busting lefts every now and then, scratching an itch or two, raising a couple of kids, and teaching folks of all ages about engineering, systems, and the systems mindset.

Just in case we are not speaking the same language, allow me to translate:

Blazing trails = being on the go and inspiring others

Busting a left = to go against the grain or do things differently

Scratching an itch = to follow your passions, enjoy life, and do the things you love

And so on…

I've been upstanding and I've been rebellious.

I've been shy and reserved and I've sought attention.

I've been in many fun clubs but also among "the wrong crowds."

I've been at the top of my class, and I've been asleep in class.

I've studied hard to do well, and I've glanced over
at someone's paper to borrow an answer.

I've been recognized for small acts of kindness,
but I've also been in big forms of trouble.

And sometimes, many of these have occurred at the same time!

Here's the reality from my perspective: "Stuff" will happen- planned, expected, or not. That's life! Some things you may enjoy, while others you may want to skip. Some things will make you feel good and proud, yet others may leave you feeling left out or embarrassed. Some friendships will thrive, and others will lapse. There are some things you will want to repeat, but there will also be things you may wish you could forget. However, one of the most important things to remember is to always choose learning. Mistakes, successes, and failures all have something in common- the opportunity to learn and grow.

There are two things that have kept me somewhat sane along my journey- one is a dream of experiencing just about everything. Two is a commitment to autonomy or dare I say "control" of my own destiny! My parents, family, teachers, and friends blessed me with guidance and experiences [good and not so good] but as it turns out, the algorithm for happiness and fulfillment is so complex that only you can discover the best ending state for you. This discovery does take time and intentional effort.

Did you notice that I chose the word "best" and not "perfect" to describe your outcomes? Perfection is one of those things that may exist [I haven't seen it]. It requires waste to achieve. Maybe it's a waste in time, or money... effort, or opportunity but I guarantee there is waste in the pursuit of perfection. Now, your parents or guardians may be hoovering over your shoulder and reading along, tempted to discard this text by this point but I assure you that my rationale is carefully grounded and sound [enough].

I pinky promise not to intentionally mislead or misinform you at any point in this book. Keep in mind that I wholeheartedly promote that you do your best and accept that this may result in winning but it could also result in a lost. I believe that when pursuing something you should apply due diligence to understand what needs to be done and to put forth sufficient effort to get it done. The result will be what it needs to be, not necessarily what you want it to be.

I believe that the true value is along the journey, not necessarily in the outcome. See, the outcome is so static, absolute even… it is what it is. But the journey is a collection of twists and turns, questions and choices, conflicts, and perspectives. It is so dynamic that no two journeys are the same and even if repeated in the exact same way, could render a variety of different outcomes.

So that is sort of what has brought us here. I am an engineer with a doctorate degree, practicing in worlds of complex systems- natural systems, engineered systems, abstract systems and some hybrid versions of such. Don't worry about what any of that means just yet! It just felt like a good time to throw some terminology at you.

I have consciously chosen to avoid the familiar notion of "common sense" for most of my life because this approach is laced with deception, boundaries, and assumptions. Instead, I have adopted a lens capable of seeing things in infinite ways. I have witnessed the dynamics of a variety of different systems and their interactions with other systems and their environment. By now, you've had a chance to at least recognize that there are multiple dimensions of our existence, including your day-to-day life and the digital worlds we sometimes opt into. This view is a great start!

Let's pause to consider a common misunderstanding before we jump too far ahead:

MTYH: You need to be an engineer to achieve the best outcomes! [Can you sense my mother/teacher voice trying to tell you what to do in this statement?]

TRUTH: Engineering and engineers are significant in the conceptualization, realization, and sustainment of engineered systems. They work among teams of talented people to evolve our world and its tools. Thus, developing an understanding of engineering, systems and developing a systems mindset will render you more effective in just about anything.

Now that we've cleared that up, here's my sales pitch… engineering is not new! It has been hammered into the heads of intellectuals and mechanically-inclined gurus near and far for thousands of years.

However, engineering awareness, and moreso an engineering systems mindset is what I am offering through this book. You will cultivate an engineering systems mindset, learning to analyze systems—their structures, behaviors, and interdependencies—to optimize your approach to problem-solving, decision-making, and the pursuit of desired outcomes. Like many professionals in the engineering systems space, you have already spent years finding and refining your crafts- whatever that is! As you learn more, you may wonder how aspects of the things you love and are passionate about work…how it works with other things…how it comes to exist or to no longer exist… This varying abstraction is laced throughout the engineering systems toolkit. My goal is to awaken you innate ability to see, influence, use and think systems.

But wait, don't say what you may be thinking! Too often I have heard the reaction, "who cares about systems"? Well…everyone should. You may not appreciate this fact yet, and likely because you may not quite understand what a system is. While I have chosen to reserve the introduction to systems for later in this book, I will tell you that almost everything exists as a component of something else. This collection of components is a part of something greater, and aim to fulfill some purpose. You are a part of your family, your social circles, clubs, and school activities, for example. You are a part of systems. Your devices have features and parts that you believe you cannot live without, like your phone's camera, microphone and screen. These components were carefully designed to form that device [or system] you love so much. Amazing, right? What about those social media challenges you have been following on online platforms. Look at you… all this time you have been using systems!

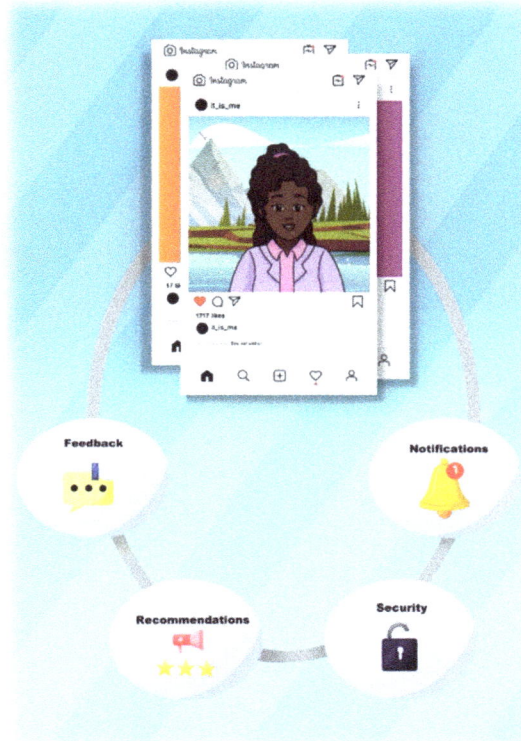

In my work with younger kids, I promote a simple affirmation as a reminder to **SEE – INFLUENCE – USE – THINK** systems throughout your daily routines. It may seem juvenile right now but throughout my life I have found tremendous value in similar things. This is why I have chosen to include a quote from a number of influential people at the beginning of each chapter of this book. Strong quotes have helped me win many awards, contests, scholarships and grants throughout my life so don't be too quick to dismiss their value.

I challenge you to try memorizing this pledge-of-sorts and recite it in your head or aloud anytime you find a need. There is actually a musical version you can download online and save to your playlist. Shoot, I'm sensing a need right now!

Let's recite it:

I see systems, systems are all around.

I influence systems, my powers are profound.

I use systems, somehow every day.

I think systems, to craft a better way.

If you choose to continue this text, I aim to take you on an adventure. You may laugh a little. You may cry a little. You may get a bit perplexed. You may feel motivated to take an action. You may find yourself questioning things you have always thought were "normal" or "just the way things are". You may feel the need to challenge something I say, and that is highly encouraged as well. I do not claim to have all the answers, any of the answers really. Instead, my hope is that you discover that there is immense power in your teenage or young adult shell. I wish that you become more aware of some of the most trivial and accessible ways of thinking, processing and doing things that can lead to an elevated view of yourself, your environment and your future.

My moment happened much later in life…but gratefully, it did happen! More insight into all of that is featured in a different book.

I challenge you read this book carefully and in its entirety. Accept this first step to **UNLEASH YOUR ENGINEERING SYSTEMS POWER!** Whatever you choose to do with what is gained, remains your choice.

Structure of the Book

Each chapter begins with an enlightening quote to help frame the upcoming narrative. Many of the quotes are like what we call heuristics in systems language. Quotes like this have improved my understanding of my experience and influenced my desire to take my journey as I have. There's enough information provided to look up the rest of the contributor's work on your own if you choose.

You will then be flooded with the self-proclaimed "brilliance" I offer related to the topics covered in each chapter. I do warn you that I have been told that I talk too much; though I promise that I have prepared the "condensed" version just for you. Try not to complain too much because there is a much longer version available for you as well!

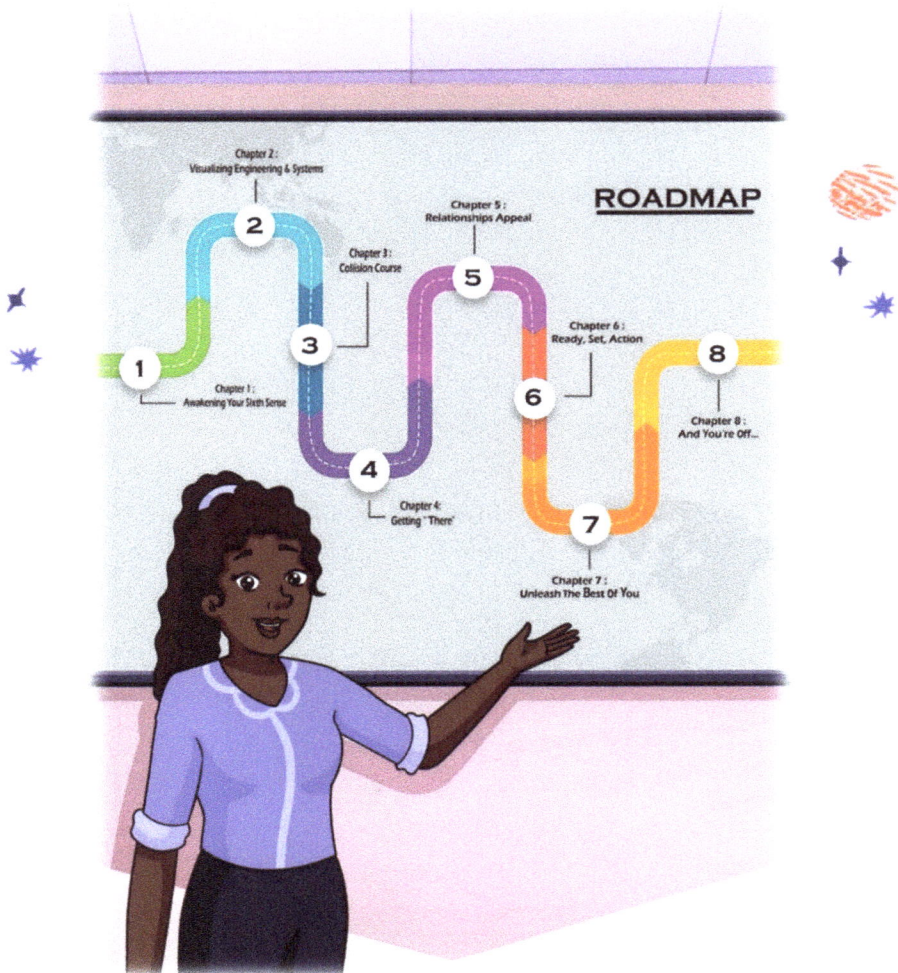

Chapter 1: Awakening Your Sixth Sense- This chapter examines how you really see this world, your immediate surroundings, the universe, metaverse and beyond. Have you witnessed certain mindsets and processes all around you? While this chapter does not dive deeply into the key question of what engineering is per se, it will begin to explore systems phenomena that you have likely encountered.

Chapter 2: Visualizing Engineering & Systems- Here, you will deep dive into engineering systems to examine what you already know about engineering and determine whether the myths that are interwoven into your beliefs (or ideology) to-date are indeed valid. You will also enhance your understanding of systems, evaluate your aptitude for engineering systems, and explore the diverse professional contexts and organizations that employ engineers.

Chapter 3: Collision Course- Lots of times, people are led to believe that they must choose one craft over another, but this limiting position is simply untrue. This chapter helps to unlock ways your journey can be "engineered" to leverage your interests and skills. There are many domains of engineering, fields that interact with engineering and applications of engineering that it's near impossible to not forge an intersection point. Through various descriptions, you will discover the relevance of engineering and the systems mindset to your own interests, hobbies and dreams.

Chapter 4: Getting "There"- So how does one become an engineer? How can you increase your exposure to engineered systems? The first step is to understand that there is more than one way to get to wherever you are going on this journey. This chapter helps you to develop the knowledge, skills and abilities needed for engineering systems- organically or systematically. Through conscious skills development exercises, you will learn to identify gaps and solve complex problems in a range of contexts.

Chapter 5: Relationships Appeal- At the time of writing this text, there is major change occurring in the social fabric of this world. Still, we must take some time to recognize social influences along our journey. How you value the relationships in your life will impact your journey. Your family, friends, teachers and others have affected who you have become, how you view yourself, and how you see the world around you. Your experiences in extracurricular activities and other organized groups have contributed to your views of collaboration, teamwork and mentorship. This chapter encourages you to explore how you perceive these relationships and experiences, whether good, bad, ugly or indifferent.

Chapter 6: Ready, Set, Action- By this point, you may be slightly overwhelmed with all of the talking and little "real" action. You're in luck! There are things that you can do now to heighten your framing, clear your blurred lenses, and get started on your refined journey. This chapter discusses activities, resources, tools, and experiences that are available in your community, on travel, online and in other virtual spaces. No matter your price point, from zero to a bazillion dollars, there are opportunities for you. Remove any temptations to focus on what may not be available because we will always be faced with constraints... and take whatever is available and make it work for you. Like the Bishop of Geneva said way back in the 1500s, "Bloom where you are planted."

Chapter 7: Unleash the Best of You- It can be rather attractive to play it safe and remain confined within your comfort zone. This is an approach taken by many, with varying results. This chapter is grounded on the idea that fear and confidence should never determine whether you try new things. Are these ambitions truly out of your reach or unrealistic? Probably NOT. Risks are inherent in everything, to one degree or another. After reading Chapter 7, I want you to know, without a doubt, that you can be anything you choose, including an engineering-oriented, systems-thinker who dares to dream big.

Chapter 8: And You're Off... After seven chapters of consciously becoming more self-aware and heightening your ability to develop and employ your engineering systems power, you are steps ahead of where you started. This chapter provides the last structured plea to refine your newfound awareness, motivation and engineering systems toolkit as you continue along your journey.

So here we go... future engineering systems enthusiast, you have officially begun to unleash your engineering systems power.

Prompting Inquiry

Distinction. What comes to mind when you think of the term "engineering"? What about "systems"? Why do these particular things come to mind? What experiences or knowledge have shaped your understanding of these terms?

Now that you've thought about each term separately, do you see a connection between engineering and systems? What kind of relationship do you think they share?

Say What? Choose a saying, quote, or heuristic that you find personally relevant. *Note: An example could be something like, "The early bird gets the worm" or "Closed mouths don't get fed".* For clarification among the terms, refer to the table.

	Saying	Quote	Heuristic
Origin	Folk wisdom, often anonymous	Attributed to a specific person	Mental shortcut, often unconscious
Form	Short, common expression	Direct excerpt from a longer work	Rule of thumb, mental shortcut
Purpose	Convey general truth or advice	Support an argument, illustrate a point	Simplify decision-making, problem-solving
Formality	Informal	Can be formal or informal	Often informal

Then address the following:

- What is the saying, quote, or heuristic?

- Describe a personal experience where you've used it (or a hypothetical situation where you might).

- How do you anticipate using it in the future?

Chapter 1
Awakening your sixth sense

But more especially and far above and beyond this, is a realm of true freedom: in thought and dream, fantasy and imagination; in gift, aptitude, and genius— all possible manner of difference, topped with freedom of soul to do and be, and freedom of thought to give to a world and build into it, all wealth of inborn individuality.

-W.E.B. DuBois

Key Topics

- Are your eyes *really* open- e.g., exploring the community, the universe, the metaverse, and everything in between?

- How does experiences, exposure, and trying new things impact you?

- How can understanding mindsets and processes influence your journey?

Wake up! The Universe is calling. Or is it the metaverse? Or something else altogether? The world is changing at inconceivable speeds, making what seems impossible today, even more probable tomorrow.

Let's start this chapter with a few questions:

Do you understand your position? Have you ever really thought about the layers of existence surrounding you? Do you think about the intricate parts that make up the things you do and see?

Every day, you navigate a world filled with people, communities, and both physical and digital phenomena. We will begin with something you encounter daily: community.

So, what exactly *is* a community? It's more than just the houses on your street or the people in your town. A community is a network of connections—people who come together, share experiences, and support one another. It's the places where we learn, grow, and connect, from the local park to the public library, to the school you attend. It's a diverse mix of individuals, each bringing unique skills and perspectives, creating something bigger than the sum of its parts. And it's a feeling of belonging, a sense of shared purpose that makes life richer and more meaningful. Your community is a vital part of your world, and you have a real stake in shaping its future. Your ideas, energy, and enthusiasm can have a ripple effect, contributing to the complex and dynamic system that makes up your community.

Our World

Our world, this vibrant blue marble we call home, isn't just a backdrop; it's a living, breathing entity, and we're all active participants in its story. From the smallest microbe to the largest whale, every living thing plays a role in the intricate dance of life on Earth. Think of it: continents drift, oceans churn, and climates shift, all while billions of people, each with unique cultures, languages, and traditions, contribute to the richness of human experience. We're a diverse species, sharing this planet with a breathtaking array of plants and animals, from the depths of the ocean to the peaks of the highest mountains.

This world is a place of stunning natural wonders. Majestic mountain ranges pierce the sky, vast oceans stretch to the horizon, and lush forests teem with life. But it's not just about the beautiful scenery. We've also built incredible cities, centers of innovation and connection, powered by technologies that allow us to communicate across vast distances and explore the world in ways our ancestors could only dream of.

But our world isn't without its challenges. We face issues like climate change, pollution, and inequality, problems that demand our attention and action. Yet, within these challenges lie incredible opportunities for positive change. And that's where you come in. You're not just a passenger on this journey; you're a driver. Your choices, your actions, your ideas—they all have the potential to shape the future of our world.

Can you think of ways you already interact with the world?

The Metaverse

Imagine stepping into a vast digital universe that feels more alive and interactive than anything you've experienced online. Welcome to the metaverse! This isn't just another corner of the internet; it's an immersive world where you can dive in and experience things as if they were right in front of you.

In this exciting realm, you can create your own avatar—a digital version of yourself that represents you in the metaverse. With your avatar, you can explore, play, learn, and connect with others in ways that go beyond the limits of the physical world. Picture it as a giant video game with endless levels, adventures, and experiences to enjoy. Whether you want to jump into an epic quest, attend a virtual concert, or simply hang out with friends, the opportunities are limitless.

One of the coolest features of the metaverse is the ability to hang out with your friends, no matter where they are in the real world. You can chat, play games, or go on thrilling adventures together, all while feeling like you're in the same room. It transforms how you socialize, making friendships feel more dynamic and engaging.

But the metaverse isn't just about fun and games. It's also a powerful platform for learning and working. You can attend virtual classes that make learning feel interactive and engaging, explore historical sites from the comfort of your home, or even collaborate with colleagues in a virtual office setting. Imagine brainstorming ideas with people from all over the globe, all while wearing your favorite hoodie!

Just like in the real world, the metaverse has its own economy. People buy and sell virtual goods, create and sell digital art, and even build businesses offering virtual services. You can turn your creativity into a source of income by designing unique virtual spaces, developing games, or crafting digital art. The metaverse is a place where your imagination can truly come to life.

One of the most exciting aspects of this digital universe is that you can create almost anything you can dream of. Whether you want to design a fantastical world, build a game, or showcase your art, the metaverse offers a platform for your creativity to shine. It's a digital playground where your ideas can flourish without limits.

While the metaverse is an incredible space filled with opportunities, it's essential to be aware of some challenges. Issues like privacy and security matter because you're sharing personal information in this digital environment. Additionally, it's important to strike a balance; spending too much time in the metaverse can affect your real-life relationships and well-being. Remember, it's all about moderation!

In a nutshell, the metaverse is an expansive and thrilling digital playground where you can explore, learn, socialize, and create. It blends elements of virtual reality, augmented reality, and the internet, paving the way for a future where you can be whoever you want and pursue whatever dreams you have.

And Beyond

Beyond our world lies the vast, awe-inspiring universe. Imagine a cosmic playground of unimaginable scale, filled with billions of stars grouped together in galaxies, swirling islands of light across the vastness of space. Our own galaxy, the Milky Way, is just one tiny speck in this grand cosmic design. Around these stars orbit planets, some with moons, each a unique world with its own story to tell.

Space, the seemingly empty canvas between these celestial bodies, isn't truly empty at all. It's filled with invisible forces like gravity, and mysterious substances like dark matter and dark energy, all working together to hold the universe together. We peer into this cosmic ocean with telescopes and send probes to explore its depths, constantly discovering new wonders and unraveling ancient mysteries. From the fiery birth of stars in nebulas to the enigmatic pull of black holes, the universe is a place of constant change and breathtaking beauty.

It's humbling to realize that everything on Earth, including you, is made of stardust, the remnants of exploded stars from billions of years ago. In a very real sense, we're all connected to the universe, part of a cosmic story that stretches back to the beginning of time. So, as you explore the world around you, remember that you're not just a resident of Earth; you're a citizen of the universe, with the power to make a difference both here and beyond.

What thoughts emerge regarding the possibility of influencing the universe?

Experiences & Exposure

It is critical to acknowledge that every person's journey is filled with unique experiences that make them whomever they've become. Sometimes it's harder to notice when your experiences have been limited to a single community or your local area but even then, we are all on a unique path.

I have included a list of common activities and milestones you may have encountered or may wish to experience. Give yourself 1 point for each activity you have experienced. Then, add those points. If you have a chance, ask your friends and adults in your life what their sum is to learn more about their journey.

I have also included a space for you to check activities you may be interested in experiencing. These are things you can continue to pursue over time.

I HAVE		I WILL
_____	Spent time with people from a different background.	_____
_____	Celebrated a cultural or religious holiday.	_____
_____	Attended a professional sporting event.	_____
_____	Gone on a vacation with family or friends.	_____
_____	Earned an "A" on a report card.	_____
_____	Engaged in a creative art form.	_____
_____	Learned to drive or got a learner's permit.	_____
_____	Volunteered for a charitable cause.	_____
_____	Taken standardized tests like the SAT or ACT.	_____
_____	Played a video game like Roblox or Minecraft.	_____
_____	Learned a second language.	_____
_____	Attended a concert or festival.	_____
_____	Discussed college or career plans with an adult one-on-one.	_____
_____	Made a mistake and learned from it.	_____
_____	Gone to a theme park.	_____
_____	Used virtual reality or augmented reality.	_____
_____	Participated in a school event after school hours.	_____
_____	Visited a public library.	_____
_____	Gone to a museum.	_____
_____	Listened to an audiobook.	_____
_____	Participated in a team sport or group activity.	_____
_____	Learned to play a musical instrument.	_____
_____	Traveled to another country.	_____
_____	Prepared a meal for yourself or others.	_____

I HAVE		I WILL
_____	Learned to ride a manual bike.	_____
_____	Learned to ride an electric bike or scooter.	_____
_____	Cooked or eaten a meal from different cuisines or cultures.	_____
_____	Participated in a club or extracurricular activity.	_____
_____	Stayed at the same school an entire school year.	_____
_____	Lived with both parents in the same household.	_____
_____	Lived in a multi-generational household.	_____
_____	Opened a bank account.	_____
_____	Gone to an aquarium.	_____
_____	Worked an internship.	_____
_____	Worked a part-time job.	_____
_____	Used artificial intelligence to learn something new.	_____
_____	Had a party to celebrate your birthday.	_____
_____	Traveled to a different place by plane.	_____
_____	Learned to swim.	_____
_____	Practiced meditation or mindfulness.	_____
_____	Spent the night in a hotel room.	_____
_____	Earned a positive recognition at school.	_____
_____	Took a financial literacy course.	_____
_____	Set a personal goal and took steps to achieve that goal.	_____
_____	Rode a horse.	_____
_____	Used your voice to speak up for your ideas/opinion.	_____
_____	Had a positive interaction with a police officer.	_____
_____	Spent time in nature, hiking, camping, or fishing.	_____
_____	Taken a college course.	_____
_____	Taken a college/university tour.	_____

Sum of "I HAVE" =

No matter your total, you are off to an amazing start!

The point of the exercise was to highlight that there's a lot out here. Life can be a wild ride, a mix of awesome highs, crushing lows, and a whole lot of "meh" in between. Every single experience, whether it's acing a test, getting your heart broken, or just chilling with friends, shapes who you become. Think of it like leveling up in a game – each challenge, big or small, gives you experience points. The tough stuff, the stuff that makes you cringe or cry, often teaches you the most. Those struggles can be the very things that make you stronger, wiser, and more resilient. They're like secret codes to unlocking hidden potential you never knew you had.

And it's not just the big moments. Everyday exposure to new ideas, different people, and unfamiliar places broadens your horizons and helps you understand the world – and yourself – better.

It sparks your imagination, ignites your passions, and gives you the raw material to build your dreams. Growing up is all about figuring things out, trying stuff on for size, and discovering what truly makes you tick. So, embrace the journey, the messy, beautiful, and sometimes awkward process of becoming you. Don't be afraid to stumble, learn from your mistakes, and keep dreaming big. You might be surprised at what you're capable of.

Can you think of any experiences that have positively affected you? Take a moment to reflect.

What is the value in your experiences and exposure?

Similarly, can you think of any experiences that have negatively affected you? Take a moment to reflect.

Don't hesitate to use your incredible library of experiences, good, bad, and everything in between. Because here's the thing: it's not just *having* those experiences that counts, it's how you *use* them. Think of your mind as the control panel for your life. You can have all the best tools and resources, but if you don't know how to use them, they won't do you much good. That's where mindset comes in. It's the lens through which you see the world, the story you tell yourself about what's possible, and the fuel that drives your actions.

Your mindset is one of your most valuable tools. It can be the difference between seeing a challenge as a roadblock or an opportunity. It can determine whether you give up when things get tough or dig in and find a way to overcome. It can even shape your ambitions and ultimately, your outcomes. By learning to understand and manage your mindset, you can take control of your growth, development, and future. It's about shifting your thinking so that you're not just reacting to what happens to you, but actively creating the life you want.

Mindset as a Tool

A mindset is a set of attitudes, beliefs, and thought patterns that shape how you perceive and approach various aspects of life. It's like a mental framework or lens through which you interpret and respond to situations, challenges, and opportunities. Mindsets can influence your behavior, decision-making, and overall outlook on life. They are not fixed traits and can be developed and changed over time with awareness and effort. There are different types of mindsets, and they play a significant role in personal growth, success, and well-being.

Four such mindsets really stand out for our purposes in this book- a curious mindset, positive mindset, problem-solving mindset, and growth mindset. What is your interpretation of each mindset?

The Curious Mindset

- Ever wonder why some people seem to breeze through challenges while others get stuck?

- What's the secret ingredient that helps some people not just survive, but *thrive*?

A big part of it is cultivating a curious mindset. This means embracing learning with a sense of wonder, approaching the world with open eyes and a thirst for knowledge. A curious mind asks questions, seeks answers, and fuels a desire to explore, understand, and grow. It's about digging deeper, whether you're fascinated by a scientific concept, captivated by a piece of art, or eager to learn about a different culture. Crucially, a curious mindset remains open to new experiences, recognizing that even the tough times have something to teach. It allows you to see obstacles not as roadblocks, but as opportunities for growth, turning challenges into valuable lessons on your journey.

The Positive Mindset

- Ever notice how some people seem to find a silver lining even in the darkest clouds?

- How do they manage to stay hopeful when things get tough?

A lot of it comes down to cultivating a positive mindset. This isn't about ignoring the tough stuff or pretending everything is perfect. Instead, it's about actively focusing on the bright side, recognizing the good in situations, and believing in your ability to overcome challenges. A positive mindset is about maintaining hope and optimism, which can be like a superpower when it comes to bouncing back from setbacks. And a key ingredient in building this kind of resilience is practicing gratitude. Regularly reflecting on the things you're thankful for, big or small, can shift your perspective and help you maintain a more positive outlook, even when life throws you curveballs.

The Problem-Solving Mindset

- Do you find yourself seeking clear-cut answers and solutions immediately?

- Do you find yourself facing seemingly impossible challenges?

A problem-solving mindset can be your key to unlocking solutions. This isn't about wishing problems away, but about viewing them as puzzles waiting to be solved, rather than insurmountable barriers. A resourceful problem-solver approaches challenges head-on, confident that a solution exists. And a crucial part of this is embracing creativity. Instead of just looking for the obvious answer, a problem-solving mindset encourages you to think outside the box, brainstorm multiple solutions, and approach problems with curiosity. It's about getting creative, exploring different angles, and not being afraid to try something new. This approach transforms obstacles into opportunities for innovation and growth.

The Growth Mindset

- Ever wonder why some people seem to constantly improve and learn, while others feel stuck?

- What's the secret to unlocking your full potential?

A lot of it boils down to having a growth mindset. This is the belief that your abilities and intelligence aren't fixed, but can be developed through dedication and hard work. It's about understanding that you're not born with a certain set of skills; you can learn and grow with effort. This perspective fosters resilience, a love for learning, and a willingness to embrace challenges. And a key part of the growth mindset is how you view failure. Instead of seeing it as a final destination, you see it as a stepping stone. It's about analyzing what went wrong, extracting the valuable lessons, and using that knowledge to improve and move forward.

Based on your understanding, gauge your proficiency level for each mindset. I'm using the levels of being "Aware", an "Active Practitioner" or an "Expert" to describe the levels. To better understand what I mean by these levels, think of mastering a mindset like learning a new skill. **Awareness** is like hearing about basketball for the first time – you understand the basic concept, but you've never actually played. **Practitioner** is like playing regularly – you know the rules, you've practiced, and you're getting better, but you're still working on your game. **Expert** is like being a pro basketball player – you've mastered the fundamentals, you can adapt to any situation, and you instinctively make the right moves. So, with mindsets, awareness means you understand the concept, practitioner means you're actively trying to use it in your life, and expert means it's become a natural part of how you think and act.

Circle a proficiency level for each mindset:

Your curious mindset

Awareness-------Practitioner-------Expert

Your positive mindset

Awareness-------Practitioner-------Expert

Your problem-solving mindset

Awareness-------Practitioner-------Expert

Your growth mindset

Awareness-------Practitioner-------Expert

Building powerful mindsets is a lifelong journey, not a destination. It's like leveling up in a game – you're constantly learning, growing, and refining your skills. Simply being *aware* of these mindsets – curious, positive, problem-solving, and growth-oriented – is a great first step. But awareness alone isn't enough. To truly empower your future, you need to become a *practitioner*. This means actively applying these mindsets in your daily life. Seek out new experiences, ask questions, embrace challenges, and celebrate your successes (and learn from your setbacks). The more you practice, the more these mindsets become ingrained habits, shaping how you approach the world. Eventually, with consistent effort, you can reach the *expert* level, where these powerful ways of thinking become second nature, guiding your decisions and actions instinctively. Remember, building strong mindsets is an ongoing process. Continuously challenge yourself, reflect on your experiences, and stay open to learning and growth. The more you invest in developing your mindsets, the more you'll unlock your potential and create the future you dream of.

Getting to Know Processes

The last thing I want to cover in this chapter are processes. Have you ever felt overwhelmed by a task, wishing there was a simpler way to tackle it? It's tempting to avoid simplification, to reduce a complex challenge to something easily digestible. Too often, though, I've seen people struggle or fail because they didn't consider the actual steps needed to achieve their goals. Truly understanding the *process*—the specific sequence of actions and dependencies—can be the key to unlocking your potential!

What is a Process?

Put simply, a process is simply a series of steps or actions you follow to complete a task or reach a goal. Think of it as your personal roadmap that guides you from where you are to where you want to be. Processes help us stay organized and efficient, providing a clear path to achieve our desired outcomes.

Just like a recipe for baking your favorite cookies or the rules for playing a board game, many everyday tasks have their own processes. Whether you're doing homework, cleaning your room, or mastering a new skill, having a clear process makes everything easier.

Why Understanding Processes Matters

Ever feel like you're just going through the motions, reacting to things as they happen, instead of being in control? Or maybe you've started something, only to realize halfway through that you missed a crucial step or should have opted for an alternative approach?

Understanding processes – how things work, step by step – can be a total game-changer. Think of it like having a roadmap instead of just wandering around hoping to reach your destination. Knowing the steps involved in anything, from studying for a test to planning a party, helps you stay organized, save time, avoid mistakes, and ultimately, achieve your goals. It reduces confusion, keeps you on track, and makes success feel much more attainable. My background in Industrial Engineering is all about this. We have a mantra: "We make things better – faster, better, cheaper." It's not just about factories and machines; it's a way of thinking that can be applied to *anything*. By breaking down complex tasks into smaller, manageable steps (or reductionism), you can identify areas for improvement and become more efficient in all aspects of your life. Understanding processes empowers you to take control, work smarter, not harder, and make the most of your time and energy. It's a skill that will benefit you immensely, not just now, but throughout your entire life.

Let's break down a familiar task, like making a video for social media, into a process. We will do so at a very high level because there are many steps we could further break down within each of these steps captured. So, at a very high-level, imagine the steps you need to follow to post a video on social media:

1. Plan Your Content: Decide what you want your video to be about.

2. Record the Video: Use your device to capture the footage.

3. Edit the Video: Add effects, music, and transitions to make it pop.

4. Write a Caption: Craft an engaging caption to attract viewers.

5. Post the Video: Share it with your followers and engage with comments.

To visualize a process, you can create a flow chart or process map. Think of it as a treasure map that leads you to your goal—except instead of treasure, you're uncovering the steps needed to complete a task. These diagrams highlight each step in a clear and organized way, helping you grasp what to do next. By laying out these steps visually, you can understand the process more easily, making each task feel more manageable.

Here's a breakdown of the shapes and symbols we'll use to show you how to illustrate the social media video creation process.

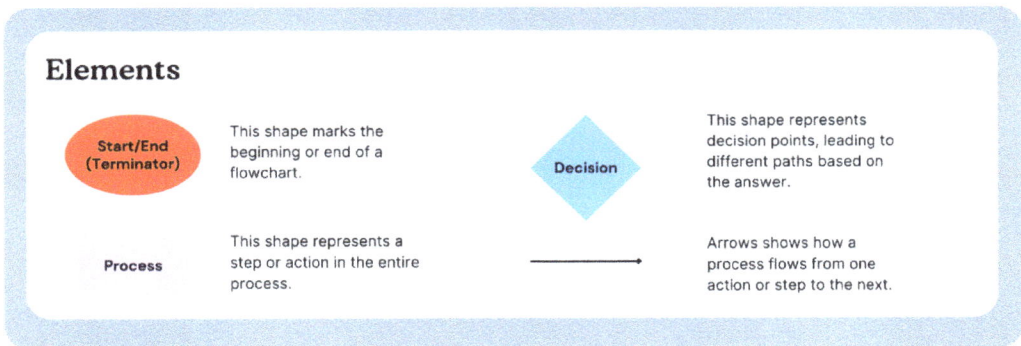

Elements

Start/End (Terminator) — This shape marks the beginning or end of a flowchart.

Decision — This shape represents decision points, leading to different paths based on the answer.

Process — This shape represents a step or action in the entire process.

Arrows shows how a process flows from one action or step to the next.

Now, review the diagram at your own pace.

The process begins with content planning, followed by video recording. After recording, you decide whether to edit the video. If you choose to edit, you then evaluate the results. If the edit is satisfactory, you write a caption and post the video. If not, you return to the recording stage. If you choose *not* to edit after recording, you simply write the caption and post.

Was the flow relatively easy to follow? Consider how this tool might help you in your own work.

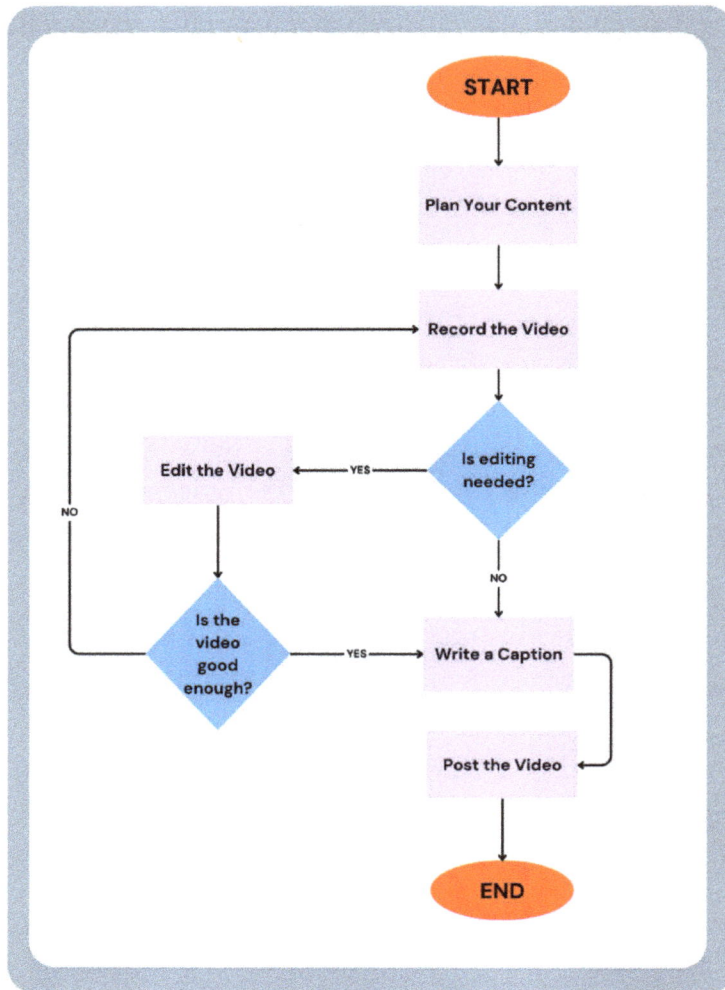

Just remember, a process is like a recipe for success in various areas of your life. It enables you to tackle tasks more effectively and efficiently. So, the next time you face a challenge, think about the process you can follow to achieve your goal. With a clear path laid out before you, you'll be well on your way to success!

Summary

As you embark on this journey through the realms of the world, the universe, and the metaverse, remember that you hold the power to shape your experiences and mindset. Your perspective can influence how you navigate life's challenges and seize opportunities. But it's not just about *what* you see, it's also about *how* you see it. Applying the right mindsets – cultivating curiosity, positivity, a problem-solving approach, and a growth mindset – will significantly impact your ability to learn, adapt, and thrive. And understanding processes – breaking down complex tasks into manageable steps – is crucial for achieving your goals. It's about knowing not just *what* you want to achieve, but also *how* to get there.

So, embrace the beauty and complexity of the world around you. Stay curious, remain open to learning, and recognize the interconnectedness of your experiences. The more you understand your position in this vast landscape, and the steps required to achieve the things you aim for, the more equipped you'll be to make meaningful contributions to your community, engage with the universe, and explore the digital frontiers.

Your journey is just beginning, and every choice you make can lead to new horizons. So, awaken your sixth sense—see the world with fresh eyes, and get ready to embark on an adventure that will shape your future!

Prompting Inquiry

Map it Out. Think about the various tasks you perform regularly. Select one activity, such as practicing a musical instrument, completing a specific chore, or working on a section of a larger project. Develop a detailed flowchart that maps out the entire process.

Be sure to include:

- All the individual steps involved.

- Any decision points where you have to make a choice.

- Potential alternative paths depending on those choices.

- The inputs and outputs of each step (what you need to start, what you produce).

- Any feedback loops or points where you might have to repeat a step.

Aiming with Intention. Pick a goal you'd like to achieve. Research the steps needed to reach it. For each step, consider the resources you'll need, the time it will take, and most importantly, how your mindsets (curiosity, positivity, problem-solving, growth) will help you succeed.

- How will you use these mindsets to overcome challenges and learn from setbacks?

- Create a plan that incorporates both the practical steps and the mental strategies you'll use.

Chapter 2

Visualizing Engineering & Systems

The ideal engineer is a composite ... He is not a scientist, he is not a mathematician, he is not a sociologist or a writer; but he may use the knowledge and techniques of any or all of these disciplines in solving engineering problems.

-Nathan W. Dougherty

Key Topics

- What do you think you know about engineering?

- What should you really know about engineering and systems?

  - Who becomes engineers?

  - What do engineers do?

  - Where do engineers work?

Engineering: the word conjures images of hard hats, complex equations, and groundbreaking inventions. While those images aren't entirely wrong, they often paint an incomplete, and sometimes misleading, picture of what engineering truly entails. Popular culture and common misconceptions have led to a number of myths about the field, from the perception that it's solely for "math geniuses" to the idea that it's a solitary pursuit. Before we dive into the reality of this dynamic and diverse profession, let's debunk some of these persistent myths and shed light on what engineering *actually* is.

Myth #1: Engineers are all math geniuses who love numbers.

- **Truth:** While math is important in engineering, it's not the only skill needed. Engineering is about applying logic, creativity, and problem-solving skills. Strong communication and teamwork are also crucial for success.

Myth #2: Engineers are introverted social outcasts.

- **Truth:** Engineers come in all personality types! Many engineering jobs require excellent communication skills to collaborate with colleagues, explain technical concepts, and present ideas.

Myth #3: Engineering is a boring and monotonous career.

- **Truth:** Engineering is a dynamic field with constant innovation and new challenges to tackle. Engineers get to work on projects that can make a real difference in the world, from designing sustainable energy solutions to developing life-saving medical devices.

Myth #4: You need to be a creative genius to be an engineer.

- **Truth:** Engineering is more about being creative within constraints. Engineers use their knowledge and ingenuity to find practical solutions to complex problems.

Myth #5: Women and minorities aren't welcome in engineering.

- **Truth:** The field of engineering is actively working towards greater diversity and inclusion. There are many inspiring women and minorities who have made significant contributions to engineering, and there are more opportunities than ever for both women and minorities to succeed in this field.

Myth #6: You need perfect grades to become an engineer.

- **Truth:** While academic performance is important, engineering schools consider a variety of factors when admitting students. A passion for engineering, relevant extracurricular activities, and a strong work ethic can all contribute to a successful application.

Myth #7: Engineering jobs are all about fixing things.

- **Truth:** While some engineers do work in maintenance and repair, many are involved in design, development, and research. They create new technologies, improve existing systems, and find innovative solutions to problems we haven't even encountered yet.

Myth #8: Engineering is a stable but low-paying career.

- **Truth:** Engineering offers excellent career prospects and competitive salaries. The demand for engineers is consistently high across various industries, and experienced engineers can earn very comfortable incomes. Furthermore, the work is often intellectually stimulating and offers opportunities for professional growth.

Myth #9: Engineering is a solitary profession.

- **Truth:** Collaboration and teamwork are essential aspects of most engineering roles. Engineers often work in teams, bringing together diverse skills and perspectives to tackle complex projects. Communication skills, both written and verbal, are crucial for collaborating effectively with colleagues, clients, and other stakeholders.

Myth #10: All engineers work in hard hats on construction sites or in labs.

- **Truth:** While some engineers do work in those environments, the field is incredibly diverse. Engineers work in a vast range of industries. They might be designing software, developing new materials, improving healthcare systems, or working on sustainable energy solutions. The possibilities are nearly endless, and many engineering jobs are office-based or involve a blend of fieldwork and office work.

If those common misconceptions aren't true, then what *is* engineering?

Turns out, engineering is the discipline, profession, and art of acquiring and applying scientific, mathematical, and economic knowledge to design, build, and improve structures, machines, devices, systems, materials, and processes. It encompasses a broad range of specializations, each with its own focus and set of challenges. Ultimately, engineering is concerned with innovation, problem-solving, and creating solutions that benefit society.

The Secret Weapon- Systems Thinking

There's a secret weapon that all great engineers use: *systems thinking*. What exactly *is* a system, though? It's not just a buzz word engineers throw around. Think of it like this: a system is a group of interacting or interdependent parts forming a complex whole. It's how things work together to achieve a common purpose.

Think about your bicycle. It's not just a bunch of individual parts scattered on the floor. The wheels, pedals, gears, brakes, and frame all work together in a specific way to create a functioning bicycle. If one part (or subsystem) is missing or broken, the whole system doesn't work as well, or maybe not at all. That's a system! And it's true for everything from a simple machine like a bicycle to something incredibly complex like the internet, a city's transportation network, or even your own body.

So many fields, including engineers are fascinated by systems. They don't just look at individual parts; they look at how those parts connect and interact, how they influence each other, how the whole system performs and even how the system may be a part in yet another system (or system of systems). Remember, the world *is* a system, so understanding how things connect and interact is valuable no matter what path you choose in life.

Types of Systems

Understanding the world around us often involves recognizing patterns and structures. These patterns and structures can be classified as different types of "systems," each with its own unique characteristics. It's important to note that this list is neither mutually exclusive nor exhaustive; it simply offers different ways to classify a system, as a single system can often fall into multiple categories. From the tangible objects we interact with daily to the complex, abstract concepts that shape our understanding, systems are fundamental to how we organize and interpret information. Below is an alphabetical list of various system types, each accompanied by a brief description to help clarify their nature and function.

- **Abstract Systems:** These systems are made of ideas, concepts, or information. Think of a mathematical equation, a computer program, or a language. They exist in the realm of thought rather than physical reality.

- **Closed Systems:** These systems don't exchange matter or energy with their surroundings. A perfectly insulated container is a theoretical example. In reality, truly closed systems are rare.

- **Complex Systems:** These systems have many interacting parts and unpredictable behavior. Weather systems, ecosystems, and the human brain are complex systems. They're often difficult to model and control because even small changes can have big effects.

- **Earth Systems:** This refers to the complex interactions between the Earth's atmosphere, hydrosphere (water), lithosphere (land), and biosphere (living things). It's the whole planet working as one huge system.

- **Enterprise Systems:** These are large, complex systems used by organizations, like businesses or governments. They handle things like finances, inventory, customer data, and communication. Think of the systems used by Amazon to process orders or a school's system for managing student records.

- **Man-made Systems:** These are systems designed and built by humans to serve a specific purpose. Cars, computers, and buildings are all man-made systems.

- **Natural Systems:** These systems exist in nature, without human intervention. Ecosystems, weather patterns, and the water cycle are examples of natural systems.

- **Open Systems:** These systems exchange matter and energy with their surroundings. Living organisms, like plants and animals, are open systems. They constantly interact with their environment.

- **Physical Systems:** Think of anything with moving parts or tangible components. A car, a bicycle, a bridge, or even the human body. These systems are made of physical objects that interact to achieve a function.

- **Service Systems:** These are systems that provide a service to people. Think of a restaurant, a library, or a hospital. They involve people, processes, and technology working together to meet a customer's needs.

- **Simple Systems:** These are systems with few parts and predictable behavior. A light switch, a lever, or a simple circuit are good examples. They're easy to understand and control.

- **Space Systems:** These are systems that operate outside of the earth's atmosphere. Satellites, the international space station, and telescopes are all examples of space systems.

- **System of Systems:** This is when multiple independent systems work together to achieve a bigger goal. Think of a city's transportation network, which includes roads, trains, and buses, or the internet, which connects countless individual computers.

Explore Your Engineering Aptitude

Now you've learned a bit about what engineering and systems really are – beyond the myths and stereotypes. Are you interested in learning more? To help you explore that, I've put together a quick checklist. Take a look at the statements below and see which ones resonate with you. There's no right or wrong answer, and checking off a few boxes doesn't automatically mean you are destined to be an engineer. This is simply a tool to spark your curiosity and help you consider whether engineering might be a field worth exploring further.

☐ Yes   ☐ No   ☐ Unsure | Do you enjoy working with numbers, formulas, and understanding how the physical world works?

☐ Yes   ☐ No   ☐ Unsure | Are you comfortable using math and science principles to solve problems and design things?

☐ Yes   ☐ No   ☐ Unsure | Do you analyze information, identify weaknesses, and come up with creative solutions?

☐ Yes   ☐ No   ☐ Unsure | Do you question things and not just accept them at face value?

☐ Yes   ☐ No   ☐ Unsure | Do you have a burning desire to understand how things work?

☐ Yes   ☐ No   ☐ Unsure | Do you enjoy taking things apart (carefully!) to see what makes them tick?

☐ Yes   ☐ No   ☐ Unsure | Are you a creative problem solver who thinks outside the box?

☐ **Yes**   ☐ **No**   ☐ **Unsure** | When faced with a challenge, do you brainstorm ideas and experiment with different approaches?

☐ **Yes**   ☐ **No**   ☐ **Unsure** | Can you clearly explain your ideas to others?

☐ **Yes**   ☐ **No**   ☐ **Unsure** | Do you listen to different perspectives and collaborate effectively with teammates?

☐ **Yes**   ☐ **No**   ☐ **Unsure** | Do you enjoy working as part of a team?

☐ **Yes**   ☐ **No**   ☐ **Unsure** | Do you see yourself taking on leadership roles, guiding projects, and motivating others?

☐ **Yes**   ☐ **No**   ☐ **Unsure** | Are you comfortable with technology and software?

☐ **Yes**   ☐ **No**   ☐ **Unsure** | Are you a lifelong learner, curious and eager to learn new things?

☐ **Yes**   ☐ **No**   ☐ **Unsure** | Do you love figuring out how things work and making them even better?

☐ **Yes**   ☐ **No**   ☐ **Unsure** | Are you interested in designing and creating electronic gadgets?

☐ **Yes**   ☐ **No**   ☐ **Unsure** | Have you ever wondered how skyscrapers or bridges are built?

☐ **Yes**   ☐ **No**   ☐ **Unsure** | Are you concerned about the environment and interested in developing sustainable solutions?

☐ **Yes**   ☐ **No**   ☐ **Unsure** | Does the idea of designing and writing code for computers, apps, and websites excite you?

☐ **Yes**   ☐ **No**   ☐ **Unsure** | Are you fascinated by robots and their potential?

☐ **Yes**   ☐ **No**   ☐ **Unsure** | Do you want to be on the front lines of solving the world's challenges, like designing earthquake-proof buildings or developing clean energy?

☐ **Yes**   ☐ **No**   ☐ **Unsure** | Does the potential of Artificial Intelligence and its applications intrigue you?

Where Engineers Work

Engineering is not as homogeneous as it may have seemed. It isn't confined to a single setting; it's a dynamic profession that thrives in a variety of environments, making it an exciting and diverse field. Think beyond the stereotypical lab coat and hard hat. Engineers are essential contributors in a multitude of workplaces, each offering unique challenges and opportunities.

From bustling tech companies developing cutting-edge software to manufacturing plants producing innovative products, engineers are the backbone of industry. They're involved in every stage of the process, from research and development and design, to production and quality control, ensuring that products and services are efficient, effective, and meet the needs of consumers.

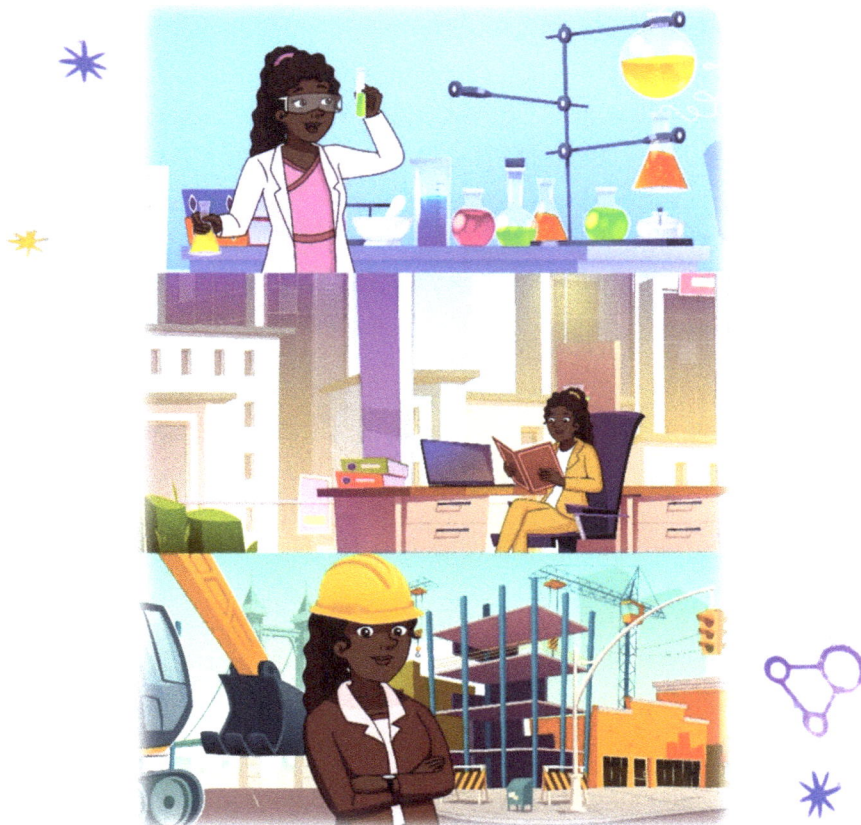

Government agencies at all levels rely on engineering expertise to design and maintain critical infrastructure, develop sustainable solutions to environmental challenges, and manage large-scale public projects. Whether it's building roads and bridges, ensuring clean water supplies, or developing renewable energy sources, engineers play a vital role in shaping our communities.

Corporations, large and small, across virtually every industry, employ engineers. From finance and healthcare to entertainment and retail, companies need engineers to design and improve their products, optimize their operations, and develop innovative solutions to stay competitive. Engineers in the corporate world might work on anything from developing new mobile apps to designing more efficient supply chains, demonstrating the broad applicability of engineering skills.

Non-profit organizations also rely heavily on engineering expertise. Engineers working in the non-profit sector might focus on developing sustainable technologies for developing countries, designing accessible infrastructure for people with disabilities, or creating innovative solutions to address climate change. These crucial projects often require navigating complex regulatory landscapes and sometimes even engaging with political processes to ensure their successful implementation. These roles often combine technical expertise with a strong social conscience, offering engineers the opportunity to make a real difference in the world.

Construction sites are where engineering designs become reality. Engineers work closely with architects, contractors, and other professionals to oversee the construction of buildings, bridges, and other structures, ensuring that projects are completed safely, on time, and within budget.

Research institutions and universities are hotbeds of innovation, where engineers push the boundaries of knowledge and explore new frontiers. They conduct research in laboratories, develop new technologies and materials, and work to solve some of the world's most pressing challenges.

Consulting firms offer engineers the opportunity to work on a variety of projects for different clients. These firms provide specialized expertise in a wide range of engineering disciplines, allowing engineers to apply their skills to diverse problems and industries.

The rise of the gig economy has also opened up new avenues for engineers. Freelancing allows engineers to work independently, choosing their own projects and clients, and offering their skills on a contract basis. This provides flexibility and autonomy, allowing engineers to explore different areas of interest and build their own businesses.

And this is just a glimpse! The reality is that engineers can be found in even more diverse settings. Some work primarily in offices, collaborating with teams, while others spend time in the field, gathering data and assessing conditions firsthand. This variety of work environments is part of what makes engineering such a dynamic and rewarding career path. But the breadth of engineering goes even further than *where* they work; it encompasses *who* they work for.

What large company comes to mind when you think of minimal or non-existent engineering involvement, and why do you think that is?

You are welcome to check their website or a job search tool to validate your assumptions.

Besides, I've prepared a list of my own. You might even recognize some of the following names, and guess what? Engineers work at every single one of them. The truth is, the list of companies that hire engineers is practically endless, spanning virtually every industry you can imagine. From tech giants to small startups, from healthcare providers to entertainment companies, engineers are essential to innovation and problem- solving across the board. So, while this list gives you a taste, it barely scratches the surface of the vast landscape of engineering opportunities.

3M	Habitat for Humanity	Philips
Airbus	Honeywell	Pizza Hut
Amazon	Instagram	Royal Caribbean
AMC Entertainment Holdings	John Deere	Samsung
American Airlines	John Hopkins University	Sephora
American Red Cross	Johnson & Johnson	Siemens
Amtrak	JP Morgan Chase	Snapchat
Apple	Keurig- Dr. Pepper	Sony
Bank of America	Kraft Heinz	SpaceX
Boeing	Lockheed Martin	Spotify
Bosch	Marriott International	St. Jude Children's Research Hospital
Central Intelligence Agency	Mc Donald's	
Chuck E. Cheese	Medtronic	Spotify
Coca Cola Company	Meta (formerly Facebook)	Stryker
Comcast	Microsoft	Target
Disney	NAACP	Tesla
Domino's Pizza	NBA	Texas Instruments
DuPont	NBC Universal	TikTok
Epic Games	Netflix	Toyota
Exxon Mobil	NFL	Unilever
FedEx	Nike	United Parcel Service
Ford	OpenAI	UnitedHealth Group
General Motors	PepsiCo	Universal
Google	Pfizer	Walmart

Career Options. Ready to explore the exciting world of engineering careers on your own? Use online resources like job search tools, to answer these questions:

- What types of engineering jobs are currently available on the job market?

- What skills and education are employers looking for?

- What types of systems and interesting projects are engineers working on today?

Who's Who? Discover the stories of remarkable engineers who have shaped our world. Research their lives and careers, paying attention to what motivated them, the challenges they overcame, and how they made a difference. Here are a few examples to inspire you:

- Mae C. Jamison
- Grace Harper
- Bill Nye
- Hedy Lamarr
- Radia Perlman
- Linus Pauling
- Emily Roebling
- Neil Armstrong
- Marie Curie
- Dr. Arati Prabhakar
- Ada Lovelace

Marvels Among Us. Identify and describe five engineering marvels, like the International Space Station (ISS), The Hoover Dam and the Burj Khalifa, Dubai. An engineering marvel is generally something technically impressive and also aesthetically pleasing, innovative, or groundbreaking in its design or scale. For each marvel, please include:

- Name and location

- Key engineering principles or innovations involved

- Approximate construction or project timeframe

- Any unique or interesting facts

SCORE

1 : 2

Chapter 3

Collision Course

A woman in harmony with her spirit is like a river flowing. She goes where she will without pretense and arrives at her destination prepared to be herself and only herself.

-Maya Angelou

Key Topics

- How can you align your engineering systems journey with your individual interests and skills?

- What are the major disciplines of engineering?

Engineering isn't a one-size-fits-all career path. The most fulfilling and impactful engineering careers are those that align with your individual strengths and passions. So, how do you discover where you fit in? The following steps will guide you through the process of tailoring your engineering journey to your unique interests and goals, helping you find the perfect niche within this exciting field.

Step 1. Unleash Your Inner Engineer

Every engineer starts with a spark – a fascination, a curiosity, a desire to create. What sparks *you*? Do you find yourself tinkering with electronics, fascinated by the inner workings of gadgets? Are you drawn to the challenge of puzzles, enjoying the thrill of logic, induction and deduction? Or perhaps you're a creative soul, happiest when designing, sketching, and bringing your ideas to life.

Think beyond the obvious. Are you a sports enthusiast? Consider how engineering could enhance performance, prevent injuries, or make sports more accessible. Even seemingly unrelated activities, like braiding hair, can spark engineering inspiration. The intricate patterns and structures can reveal underlying principles of design, materials science, and mathematics, potentially leading to breakthroughs in other fields. No matter your passion, an engineering systems mindset can help you analyze, optimize, and achieve your goals. Jot down a few of your interests in the space provided so you can begin to see how your unique interests can connect to the global challenges we face.

Step 2. Connect Interests to Global Challenges

Your interests aren't just hobbies; they're potential superpowers waiting to be unleashed on the world's biggest problems. What global challenges keep you up at night? Are you worried about the future of our planet, the health of our communities, or the opportunities available to everyone? From climate change and sustainable development to access to education and healthcare, there are countless ways to make a difference. Now, consider how your passions – whether it's music, technology, art, or even something unexpected– could be part of the solution. Imagine the possibilities! Connecting your interests to these challenges will fuel your engineering journey with purpose and meaning. As we move toward Step 3, think about these questions grounded in our current global challenges:

- What are you most curious about in the world – from climate change to artificial intelligence, or even how the human brain works? Could your curiosity be the spark for an engineering solution?

- If you could invent any gadget or technology to make life better, what would it be and which global challenge would it help address (like access to clean water, or quality education)?

- Are you a natural problem-solver? Do you enjoy figuring out how things work or finding creative solutions to tricky situations? These skills are essential for tackling complex engineering challenges.

- Imagine a future where everyone has access to clean energy and resources. What role could *you* play in making that a reality, using your unique talents and interests?

- Do you enjoy working with your hands, building things, or tinkering with technology? Hands-on skills are invaluable in engineering and can be applied to anything from building sustainable infrastructure to developing new medical devices.

- Are you passionate about a particular social issue, like environmental protection, reducing poverty, or promoting equality? Engineering can be a powerful tool for positive social change. How could you combine your passion with engineering skills?

- If you could collaborate with a team of experts to solve a global problem, what problem would you choose and what would your contribution be?

- Do you like learning new things and staying up-to-date on the latest discoveries in science and technology? A love of learning is essential for engineers, as the field is constantly evolving.

- What are your favorite subjects in school, and how could those subjects connect to the grand challenges facing our world? For example, math and science are fundamental to engineering, but even subjects like art and design can contribute to innovative solutions.

- If you could travel anywhere in the world to work on an engineering project that makes a difference, where would you go and what would you do?

- Think about your favorite sport or physical activity. How could engineering principles be used to improve performance, prevent injuries, or make sports more accessible and inclusive? Could you design better equipment, develop new training techniques using data analysis, or even create adaptive sports technology for athletes with disabilities?

Step 3. Problems Worth Solving [to you]

What problems are *you* most motivated to solve? This is where your values and passions intersect with the needs of the world. Are you driven by a desire to improve lives through better healthcare? Are you passionate about creating a more inclusive society through assistive technologies? Or are you committed to building a sustainable future by designing resilient infrastructure? Identifying the problems you want to solve is a crucial step in defining your purpose. It will guide you toward a path that not only utilizes your skills but also fulfills your deepest values.

Engineering Disciplines

You've discovered your engineering spark, connected it to global challenges, and identified the problems you're driven to solve. Now, it's time to explore the vast and exciting world of engineering disciplines. From the time-tested foundations of civil engineering, shaping the physical world around us, to the cutting-edge innovations of bioengineering, revolutionizing medicine and healthcare, the field is incredibly diverse. And it's not just about the traditional fields anymore. Driven by rapid technological advancements and evolving societal needs, new specializations are constantly emerging, opening up even more possibilities. This dynamic landscape means that no matter your interests, you're more likely than ever to find an engineering path that's the perfect fit for you.

The following pages will introduce you to a range of engineering disciplines, both established and emerging. We'll explore what each field entails, the kinds of problems engineers in these areas tackle, and how they contribute to solving global challenges. As you explore, keep your own interests and values in mind. Which fields spark your curiosity? Which problems resonate most strongly with you? This exploration is just a sampler of the disciplines across engineering.

Aerospace Engineering

Aerospace engineers design, develop, and test aircraft and spacecraft, along with the supporting technologies and systems. They work on everything from commercial airliners to satellites, applying principles of aerodynamics, propulsion, and materials science.

Agricultural Engineering

Agricultural engineers apply engineering principles to agriculture, focusing on improving efficiency, sustainability, and productivity in food production and processing. They design and develop equipment, systems, and processes for farming, irrigation, and food processing.

Architectural Engineering

Architectural engineers bridge the gap between architecture and engineering, focusing on the structural and mechanical systems that make buildings function. They design and oversee the construction of buildings, ensuring they are safe, efficient, and comfortable.

Biomedical Engineering

Biomedical engineers improve human health by applying engineering principles to the design and development of medical devices, diagnostic tools, and therapies. Their work ranges from prosthetics and pacemakers to drug delivery systems and tissue engineering.

Ceramic Engineering

Ceramic engineers develop and improve ceramic materials, understanding their unique properties to create products ranging from everyday items like glass and porcelain to high-tech applications in aerospace, electronics, and medicine.

Chemical Engineering

Chemical engineers apply principles of chemistry, physics, biology, and mathematics to design and develop processes for transforming raw materials into valuable products. These products range from fuels and pharmaceuticals to plastics, foods, and advanced materials.

Civil Engineering

Civil engineers design, construct, and maintain the physical infrastructure that supports modern society. This includes everything from roads, bridges, and buildings to water systems, transportation networks, and energy facilities.

Computer Engineering

Computer engineers bridge the gap between hardware and software. They design, develop, and test computer systems and components, from microprocessors and embedded systems to networks and software applications.

Electrical Engineering

Electrical engineers deal with the design, development, and application of systems and devices that use electricity, electronics, and electromagnetism. This broad field encompasses power generation and distribution, telecommunications, and consumer electronics.

Engineering Physics

Engineering physicists use a deep understanding of physics to solve complex engineering problems and create innovative technologies. They bridge the gap between fundamental scientific discoveries and practical engineering applications.

Environmental Engineering

Environmental engineers solve environmental challenges by applying engineering principles to the design of systems for water treatment, air pollution control, waste management, and site remediation.

Geological Engineering

Geological engineers solve engineering problems related to the Earth, including site investigation, foundation design, slope stability, and the management of natural hazards.

Industrial Engineering

Industrial engineers are problem-solvers who design, analyze, and improve systems of people, materials, information, equipment, and energy to enhance efficiency, productivity, quality, and cost-effectiveness.

Marine Engineering

Marine engineers design, build, and maintain vessels and structures that operate in the challenging marine environment, considering factors like hydrodynamics, corrosion, and weather conditions.

Materials Engineering

Materials engineers focus on the design, development, processing, and characterization of materials to create products with specific properties. They work with metals, polymers, ceramics, and composites.

Mechanical Engineering

Mechanical engineers apply principles of physics, mathematics, and materials science to design, develop, and manufacture mechanical systems. These systems range from simple tools to complex machines.

Mechatronics Engineering

Mechatronics engineers combine mechanical, electrical, computer, and control engineering principles to create innovative solutions in areas like robotics, automation, and smart devices.

Nuclear Engineering

Nuclear engineers focus on harnessing nuclear energy for power generation and developing applications of radiation in various fields, including medicine, industry, and research.

Ocean Engineering

Ocean engineers tackle the unique engineering challenges posed by the ocean environment, developing solutions for offshore platforms, underwater vehicles, coastal protection, and marine renewable energy.

Petroleum Engineering

Petroleum engineers are responsible for the efficient and sustainable recovery of oil and natural gas reserves, from exploration and drilling to production and reservoir management.

Social Engineering

Social engineers apply principles from social sciences like sociology, psychology, and economics to design and implement solutions for complex societal challenges, such as poverty, inequality, and community development.

Software Engineering

Software engineers apply engineering principles to the design, development, testing, and maintenance of software systems. They create the programs and applications that power our digital world.

Systems Engineering

Systems engineers are an interdisciplinary team who focus on the design, development, and management of complex systems, ensuring all components work together effectively to achieve a common goal.

From designing spacecraft to developing life-saving medical devices, engineering encompasses a vast array of disciplines, each offering unique challenges and opportunities. This diverse landscape allows you to find a path that aligns with your individual interests and skills, whether it's building bridges, creating software, or developing sustainable energy solutions. The fields described here represent just a snapshot of the many specializations within engineering, demonstrating the breadth and depth of this dynamic profession. Remember, the engineering journey is a continuous learning process. Stay curious, explore different disciplines, and don't be afraid to experiment and innovate to discover where your passions and talents can best contribute to shaping the future.

Prompting Inquiry

Engineering Inspiration. Immerse yourself in the world of cutting-edge engineering projects through documentaries, podcasts, or online resources. As you explore these projects, let your imagination run wild! Consider what inspires you about these projects and what you might want to explore further.

- **Capture Your Thoughts:** Keep a journal or sketchbook to record your thoughts, observations, and ideas as you learn about these projects.

- **Creative Expression:** Choose one project that particularly inspires you and express your understanding of it through a creative medium. This could be a drawing, a short story, a poem, a song, or any other form of artistic expression.

- **Share Your Inspiration:** Share your creative work with others and explain what inspired you about the project.

Aptitude Assessments. Take at least two different online quizzes or career assessments related to engineering.

- **Record Your Results:** Document the specific engineering disciplines or career paths suggested by each assessment.

- **Identify Patterns:** Are there any engineering disciplines that are consistently recommended across both assessments? Are there any significant discrepancies or conflicts between the results?

- **Self-Reflection:** Now, consider your own interests, skills, and values. For each discipline suggested by the assessments (whether consistently or conflictingly), ask yourself: Does this discipline genuinely interest me? Why or why not? Do I possess the necessary skills or aptitudes for this field? If not, are these skills I could develop? Does this career path align with my values and long-term goals?

- **Explain Your Reasoning:** For each discipline, briefly explain why you think it may or may not be a good match for you based on your self-reflection. Be specific! Don't just say "I'm good at math"; explain how your math skills would be relevant to that specific engineering discipline.

Synthesize Your Findings: Based on your analysis, which engineering disciplines seem like the best fit for you? Are there any areas you'd like to explore further?

Roadmap

Chapter 4

Getting "There"

Two roads diverged in a wood, and I-,

I took the one less traveled by, and that has made all the difference.

-Robert Frost

Key Topics

- How can you develop the knowledge, skills and abilities needed for engineering systems, both organically or purposefully?

- What are paths to becoming an engineer or engineering-minded?

You haven't yet decided that you want to become a master problem-solver, someone who can tackle complex challenges and see the connections others miss? That's OK, I can be quite persistent. This chapter is all about unlocking your potential to think like an engineer, even if you don't necessarily want to *become* one. Because here's the secret: developing the ability to identify gaps and solve problems isn't just for engineers; it's a valuable skill that will serve you well in *any* area of life.

We'll explore how you can cultivate these skills, both organically, by weaving them into your everyday routine, and more purposefully, by taking specific steps to strengthen your abilities. We'll explore the knowledge, skills, and abilities (KSAs) that form the foundation of engineering systems thinking, and how you can develop them through both natural curiosity and focused effort. And finally, we'll discover some of the many paths you can take to become an engineer, or simply to cultivate an "engineering-minded" approach that will empower you to tackle any challenge life throws your way. So, whether you dream of building bridges, designing apps, or simply becoming a more effective problem-solver, this chapter will give you the tools and insights you need to get started.

Areas of Systems Thinking Skills (ASTS)

Systems thinking is a subject of great interest to me – it's a way of looking at the world that takes practice, and just like gaming, there are specific skills involved.

Think of these skills as your systems thinking toolkit. The more tools you have, the better equipped you'll be to tackle any challenge, whether it's figuring out how to adapt to unexpected events and opponent actions or understanding why your school's lunch program isn't working as well as it could.

I've broken down these skills into 14 key areas I call the areas of systems thinking skills (ASTS), and they're all connected, just like the parts of your gaming console. Some are about how you *see* things – like Multiple Perspectives, realizing that there's more than one way to look at a problem, just like there are many tactics and strategies to your playstyle.

Others are about how you *think* – like Hypothetical and Inferential Consideration, which is like planning your next move by thinking about what *could* happen. Then there's Paradoxical and Ambiguity Tolerance – that's being okay with the fact that sometimes things aren't clear, just like sometimes a new patch or update may require you to quickly learn and master the effects of such change.

And some are about *use* – like Creativity, which is essential for coming up with new solutions to a problem when faced with a new challenge or obstacle.

In this section, I provide a brief description of each ASTS, exploring what they mean, how they show up in your life, and how you can develop them (Appendix C). Remember, these aren't just skills for engineers. They're skills for *life*. Just like knowing how to build and craft in a survival game can help you navigate the world, mastering these systems thinking skills will empower you to navigate any challenge you encounter.

ASTS. Multiple Perspectives
(or Think in Different Ways)

Imagine you and a friend are arguing about the best pizza toppings. You like pepperoni, they like veggies. You're both looking at the same pizza (the situation), but from different angles. Multiple perspectives is like that – realizing that there are lots of different ways to look at anything, not just pizza. It means understanding that other people have their own experiences, backgrounds, and opinions that shape how they see things, and that those perspectives are just as valid as yours. It's about trying to step into someone else's shoes for a moment, even if you don't agree with them.

Can you think of an example of **Multiple Perspectives** in your everyday life?

ASTS. Different Scales of Abstraction
(or Separate into Parts and Put it Together)

Imagine you're looking at a picture of your city. From far away, you see the whole thing – the shape of it, where the major roads are, maybe the big parks. That's like looking at a system from a high level of abstraction – you see the big picture. Now, zoom in. You can see your neighborhood, the houses, the trees. That's a lower level of abstraction. Zoom in even more, and you see your house, your room, your desk. That's an even lower level. Different scales of abstraction means understanding that systems can be looked at from these different levels, from the very broad overview to the tiny details, and that each level tells you something different and important.

Can you think of an example of **Different Scales of Abstraction** in your everyday life?

ASTS. Interconnections, Intrarelationships, and Dependencies
(or Explore Things that are Closely Connected)

Think about your favorite sports team. The players are all connected, right? They depend on each other to win. The quarterback needs the linemen, the receivers need the quarterback, and everyone depends on the coach. Interconnections, intrarelationships, and dependencies are like that – understanding how the different parts of a system, whether it's a team, a video game, or even your family, are connected and rely on each other. It's about seeing how one person's actions can affect everyone else, or how changing one thing in a system can have ripple effects throughout.

Can you think of an example of **Interconnections, Intrarelationships, and Dependencies** in your everyday life?

ASTS. Dynamic Behavior

(or Consider What it Does and Could Do)

Think about the weather. It's always changing, right? Sometimes it's sunny, sometimes it's rainy, sometimes it's a mix of both. That's dynamic behavior – it means that systems aren't static; they're constantly changing and reacting to different things. Just like the weather changes based on temperature, wind, and other factors, systems change based on the relationships between their different parts, what's happening around them, and even what people do. Dynamic behavior means understanding that change is normal and that systems are always adjusting and evolving.

Can you think of an example of **Dynamic Behavior** in your everyday life?

ASTS. Stock, Flow and Delay

(or Understand What is Needed, Used and the Time Required to Realize Change)

Consider your phone's battery – the battery level is the stock, how fast it drains is the flow, and there's a delay between when you plug it in and when it's fully charged. Stock, flow, and delay means understanding how these three things are related in systems. It's about recognizing that resources (like time, money, or even information) accumulate, deplete, and that there are often lags in how changes affect the system.

Can you think of an example of **Stock, Flow and Delay** in your everyday life?

ASTS. Feedback

(or Monitor the Effects of Different Actions)

Imagine you're playing a video game. You make a move, and the game reacts – maybe your character jumps, or shoots, or takes damage. That reaction is feedback – it's information that tells you whether your move was effective or not. Feedback is like that in all systems. It's how the system responds to changes, telling you whether things are working as intended or if adjustments need to be made. It can be positive (you did something well, keep doing it!) or negative (you need to change your approach).

Can you think of an example of **Feedback** in your everyday life?

ASTS. Non-linear Relationships

(or Discover How Things are Connected at a Broader Level)

Think about this: studying for an extra hour might improve your grade a little, but studying for ten extra hours might not necessarily improve it ten times as much. Non-linear relationships are about understanding that cause and effect aren't always straightforward; sometimes, small changes can have huge consequences, and big changes might not have the impact you expect.

Can you think of an example of **Non-linear Relationships** in your everyday life?

ASTS. Mental and Formal Models

(or Examine What you Think you Know and What you Need to Know)

Imagine you're craving a chocolate cake. You have a *mental model* of what that cake should be like: moist, chocolatey, maybe with some frosting. You know, generally, what ingredients go into a cake and the basic steps involved (mix stuff, bake it). This is your intuitive understanding of cake-making based on whatever you know and your assumptions. But that's different from a formal model. The formal model is the actual recipe you find online or in a cookbook. It lists the exact ingredients, the precise measurements, the oven temperature, and the specific baking time. It's the detailed, step-by-step plan you follow to turn your mental picture of a delicious cake into a reality. Your mental model is your general understanding, your vision. The formal model is the structured recipe, the specific instructions you need to make it happen.

Can you think of an example of **Mental and Formal Models** in your everyday life?

ASTS. System Structure and Boundary

(or Realize What it is and What it is Not)

Imagine a video game again. The game itself has a structure – different levels, characters, items, rules. It also has a boundary – it's contained within the console or computer. System structure and boundary are about understanding how a system is organized (structure) and what's included and what's not (boundary). A seemingly simple accessory, like a game controller, could expose a boundary issue if it works seamlessly on one platform but is incompatible with another. Understanding these nuances of structure and boundary is crucial for effective system design and implementation.

Can you think of an example of **System Structure and Boundary** in your everyday life?

ASTS. Conceptual Modeling

(or Make it Easier to See and Understand)

Conceptual modeling – creating a simplified representation of a system to make it easier to understand. It's like making a map of your neighborhood – it's not a perfect replica, but it shows the important streets and landmarks. Conceptual models can be diagrams, stories, metaphors, or even just a few key words that capture the essence of a system.

Can you think of an example of **Conceptual Modeling** in your everyday life?

ASTS. Prospection and Prediction

(or Predict What May Happen Next or Over Time)

Imagine you're playing chess. You don't just make a move without thinking about what your opponent might do next, right? You try to anticipate their moves and plan your own strategy accordingly. Prospection and prediction are like that – it's about thinking ahead and trying to figure out what might happen in the future based on what you know now. It's not about being a fortune teller; it's

about using your knowledge of how systems work to make educated guesses about potential outcomes.

Can you think of an example of **Prospection and Prediction** in your everyday life?

[ ]

ASTS. Hypothetical and Inferential Consideration

(or Gather Useful Clues from Other Things)

Imagine you're reading a mystery novel or watching a tv series. You're given some clues, and you try to infer who the culprit is, even though it's not explicitly stated. That's inferential consideration – using the information you have to draw conclusions. Hypothetical consideration is similar, but it involves thinking about "what if" scenarios. What if the butler did it? What if the victim wasn't who they seemed to be? Hypothetical and inferential consideration is about combining these two types of thinking to explore different possibilities and make informed judgments. It's like being a detective, using clues and imagination to solve a puzzle.

Can you think of an example of **Hypothetical and Inferential Consideration** in your everyday life?

[ ]

ASTS. Paradoxical and Ambiguity Tolerance

(or Accept that all the Information May Not be Available)

Ever feel like life is full of contradictions? You want to hang out with friends, but you also need to study. You're good at art, but you're told you should focus on science. That's the paradoxical nature of things – sometimes, two seemingly opposite things can both be true. Ambiguity is similar – it's when things aren't clear, when there's more than one way to interpret something, or when you don't have all the information you need. Paradoxical and ambiguity tolerance is about being comfortable with these uncertainties. It's about accepting that life isn't always black and white, that there are gray areas, and that sometimes you have to make decisions without knowing all the answers. Think about it like trying to solve a puzzle with missing pieces – you have to be okay with some uncertainty and use your best judgment to fill in the gaps.

Can you think of an example of **Paradoxical and Ambiguity Tolerance** in your everyday life?

ASTS. Creativity

(or Brainstorm or Improve Ideas and Alternatives)

Think about your favorite artist, musician, or even a friend who's really good at coming up with new ideas. Creativity is that ability to think outside the box, to come up with new and original ideas, and to find innovative solutions to problems. It's not just about art; creativity is essential in all fields, including engineering, science, business, and even everyday life. It's about seeing things in a new way and finding new ways to connect ideas.

Can you think of an example of **Creativity** in your everyday life?

Tools for Critical Thinking

Sharpening your critical thinking skills is like adding high-powered tools to your mental toolkit. It's not just about knowing facts; it's about knowing *what to do* with those facts, how to connect them, question them, and use them to solve problems creatively. This section will introduce you to several powerful frameworks for critical thinking, from analyzing systems with multiple perspectives to embracing ambiguity and fostering your creativity. Think of these frameworks as maps and compasses, guiding you through complex situations. But critical thinking isn't confined to textbooks or classrooms. Just like any skill, it's honed through practice. We'll also explore everyday ways you can organically weave critical thinking into your routine, from simply observing your surroundings more carefully to actively questioning the way things are done. Developing these skills will empower you to identify gaps, analyze complex problems, and craft effective solutions, not just in engineering, but in every aspect of your life. Remember, the more you practice, the more naturally these skills will become second nature.

The 5 Whys Method

The 5 Whys method is a simple yet powerful problem-solving technique that involves repeatedly asking "Why?" five times to drill down to the root cause of a problem. Starting with the initial problem statement, you ask "Why?" to understand the immediate cause. Then, you ask "Why?" again about that cause, and so on, repeating the process five times. Each "Why?" question should build upon the previous answer, leading you deeper into the chain of cause and effect. The idea is that by the time you've asked "Why?" five times, you'll have moved past superficial symptoms and uncovered the underlying systemic issue that needs to be addressed to prevent the problem from recurring. While the name suggests five iterations, the actual number of "Whys?" may vary depending on the complexity of the problem.

Let's explore "Why am I not interested in my coursework?" using the 5 Whys:

1. **Why am I not interested in my coursework?** Because I find it boring and irrelevant to my life.

2. **Why do I find it boring and irrelevant?** Because the material is presented in a dry and unengaging way, and I don't see how it connects to my future goals.

3. **Why is the material presented in a dry and unengaging way, and why don't I see the connection to my future goals?** Because the teaching methods are primarily lecture-based, and the curriculum doesn't offer real-world applications or examples that resonate with me. Also, I haven't taken the time to clearly define my future goals and explore how my coursework could actually help me achieve them.

4. **Why are the teaching methods primarily lecture-based, and why haven't I defined my future goals?** Because the school's resources are limited, making it difficult for teachers to implement more interactive learning experiences. Also, I haven't sought out guidance from counselors or mentors to help me explore career paths and connect them to my studies.

5. **Why are the school's resources limited, and why haven't I sought guidance?** Because the school district faces budget constraints, and I haven't been proactive in seeking out available resources like career counseling or extracurricular activities related to my interests.

This 5 Whys analysis reveals that the lack of interest in coursework stems from a combination of factors: unengaging teaching methods, a perceived lack of relevance to future goals, limited school resources, and a lack of proactive career exploration. Addressing these root causes, rather than just the surface symptom of "lack of interest," is key to finding a solution. This might involve advocating for more engaging teaching methods, seeking out real-world learning opportunities, exploring career options, or connecting with mentors.

Now you try the method for another problem: **You are consistently late to school.**

This is a personal problem where the causes are likely within your control.

1. Why am consistently late to school?_____

_____

_____

_____

2. _____

_____

_____

_____

3. _____

_____

_____

_____

4. _____

_____

_____

_____

5. _____

_____

_____

_____

Hopefully this example helped you to explore issues like your morning routine, transportation, time management, or sleep habits to discover the root cause of the problem.

Fishbone Diagram

The Fishbone Diagram, also known as the Ishikawa Diagram or Cause and Effect Diagram, is a visual tool used for brainstorming and identifying the potential root causes of a specific problem or effect. Resembling a fish skeleton, the "bones" represent major categories of potential causes, such as manpower, materials, methods, machinery, measurement, and environment (often referred to as the 6 Ms). These categories branch off the main "spine" of the diagram, which represents the problem statement. Specific causes are listed within each category, which are represented as smaller bones branching off the category bones. This structured approach helps systematically explore a wide range of potential causes, facilitating a deeper understanding of the problem and leading to more effective solutions.

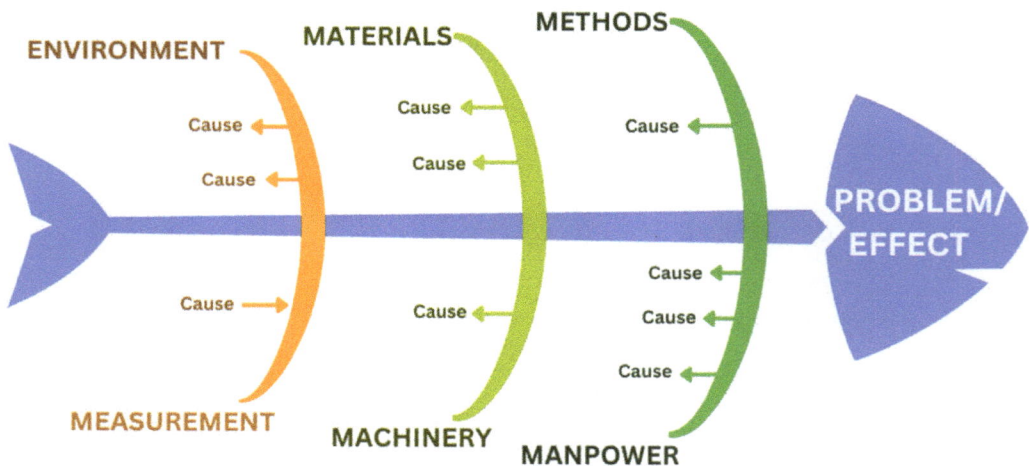

We will work on an example together.

Think about the problem of low student test scores. I will walk you through a discussion of each of the 6Ms in this context- i.e. manpower, materials, methods, machinery, measurement, and environment.

Manpower (or People): This category considers the individuals involved in the process, including teachers, students, support staff, and administrators. It looks at factors like their skills, training, experience, motivation, and workload. For low test scores, this might include teacher qualifications, student effort, or the availability of support staff.

Materials: This refers to the resources used in the educational process, such as textbooks, curriculum materials, technology, classroom supplies, and learning aids. For low test scores, this could involve outdated textbooks, insufficient technology access, or a lack of appropriate learning resources.

Methods: This encompasses the teaching strategies, instructional techniques, lesson plans, assessment methods, and overall educational approaches used. For low test scores, this might include ineffective teaching methods, a lack of differentiated instruction, or inappropriate assessment tools.

Machinery (or Equipment/Technology): This includes the physical tools and technology used in the learning process, such as computers, projectors, software, lab equipment, and other classroom technology. For low test scores, this could involve outdated computers, unreliable internet access, or a lack of necessary lab equipment.

Measurement: This category examines how student performance is evaluated, including tests, quizzes, assignments, grading policies, and other assessment methods. For low test scores, this might involve biased testing methods, a lack of formative assessment, or inconsistent grading policies.

Environment: This refers to the physical and social context in which learning takes place, including the classroom environment, school climate, home environment, peer influences, and community factors. For low test scores, this could include a noisy classroom, a lack of parental support, or negative peer pressure.

Using this understanding and any additional brainstorm you may have completed, create a fishbone diagram for this problem.

SCAMPER

The last framework we will cover is SCAMPER. SCAMPER is a creative brainstorming technique that encourages you to look at problems from a variety of angles by prompting you to consider seven different perspectives. Each letter in SCAMPER represents a different approach: **S**ubstitute (can you replace part of the problem or solution with something else?), **C**ombine (can you merge two or more parts of the problem or solution?), **A**dapt (can you modify or adjust something existing to fit your needs?), **M**odify (can you change the size, shape, color, or other attributes of something?), **P**ut to other uses (can you find a new application for something already in existence?), **E**liminate (can you remove or simplify any part of the problem or solution?), and **R**everse (can you flip, invert, or do the opposite of what's expected?). By systematically working through each of these prompts, SCAMPER helps you break free from conventional thinking and generate a wider range of innovative ideas.

As an example, think about the need to reduce food waste at home using the SCAMPER approach. How might we reduce the amount of food wasted at home? This is a complex problem with many facets, making it ripe for SCAMPER's diverse perspectives. Does any of these ideas come to mind- e.g. solutions like considering substituting ingredients, combining preservation methods, adapting packaging, modifying portion sizes, putting food scraps to other uses, eliminating unnecessary packaging, and reversing the typical food supply chain. By applying SCAMPER, you can ensure you've explored a wide range of possibilities. Feel free to continue to brainstorm.

Everyday Skill-Building

Now, let's get practical. How do you actually build these critical thinking superpowers? It's not about suddenly becoming a engineering systems guru overnight; it's about incorporating specific practices into your everyday life. One key aspect is actively seeking diverse perspectives. Talk to people from different backgrounds, disciplines, and walks of life. Their unique viewpoints can reveal aspects of a problem, or potential solutions, that you might never have considered on your own. Don't just rely on opinions; back them up with research. Gather data and information from credible sources to fully understand the context and different facets of the problem you're trying to solve.

Practice is essential. Start small by identifying gaps and inefficiencies in your daily routine. How can you optimize your tasks or solve minor household problems, for example? These everyday puzzles are great starting points. Engage your mind with logic puzzles, brain teasers, or online problem-solving games. They're a fun way to sharpen your critical thinking abilities.

Take on projects that require you to identify a need, research solutions, and develop a plan. This could be anything from organizing a community event to building a piece of furniture.

Finally, embrace failure as a learning opportunity. Don't be afraid to experiment and make mistakes. View them as valuable feedback, guiding you closer to a successful solution. Be prepared to refine and iterate on your approach. The first solution you come up with might not be the best, so be open to gathering feedback, testing your solutions, and iterating until you achieve the desired outcome. Remember, building these skills is a journey, not a destination. The more you practice, the more naturally you'll approach challenges with a critical, systems thinking, and solution-oriented mindset.

Intentional Skill Development

Beyond these everyday practices, there are more intentional ways to accelerate your development of critical thinking and engineering-related skills. Formal education is a cornerstone. Taking challenging math and science courses throughout grade school builds a strong foundation for future engineering studies. Consider advanced classes like calculus, trigonometry, physics, and chemistry to deepen your understanding.

Engineering summer programs and camps offer a fantastic opportunity to immerse yourself in the world of engineering. Many universities and organizations host these programs, designed to introduce people to engineering concepts and provide hands-on experience. They can give you a taste of different engineering disciplines and help you discover where your interests lie.

Online courses and resources have exploded in recent years, offering a wealth of information at your fingertips. Platforms like Coursera, edX, and MIT OpenCourseware provide access to both free and paid courses covering everything from fundamental engineering principles to coding, software development, and specialized engineering topics.

In addition, don't underestimate the power of mentorship. Connecting with a mentor who is an engineer can provide invaluable guidance and insights into the profession. They can answer your questions, share their real-world experiences, and offer personalized advice on how to prepare for an engineering career.

Reading books and articles about engineering can also broaden your horizons, introducing you to different fields, famous engineers, and groundbreaking innovations. This can spark your interest and fuel your passion for engineering.

By combining organic curiosity with these more purposeful learning experiences, you can build a robust foundation of knowledge, skills, and abilities that will prepare you for a successful engineering career... or simply empower you with an engineering systems mindset that can benefit you in any field.

Before we move on, reflect on a time recently when you encountered a problem, big or small. Considering what we've covered through this section, what is one specific way you could have approached that problem differently considering actively seeking a diverse perspective or by gathering more information from credible sources? What might have been a different outcome or a new solution you could have considered?

Engineering Pathways

So, you're intrigued, you're developing your skills, but now you might be wondering, "What are the actual pathways to becoming an engineer?" There are several options, and the best route for you will depend on your educational background, financial goals, and desired outcomes. Let's explore the common paths in the next section.

Traditional Four-Year Degree

The most common pathway to becoming a licensed professional engineer (PE) is the traditional four-year bachelor's degree. This involves earning a Bachelor of Science (BS) in a specific engineering discipline, such as mechanical, civil, or electrical engineering. The curriculum for these programs typically includes a robust foundation in core engineering subjects like math, physics, and chemistry, followed by fundamental engineering courses covering topics such as statics, dynamics, and circuits. Finally, you take specialized coursework tailored to your chosen discipline, providing you with the in-depth knowledge and skills needed for professional practice. This well-rounded education provides a strong base in engineering principles and prepares you for the professional licensing exams required to become a PE.

Two-Year Associate's Degree followed by Bachelor's Degree

Another common path to an engineering degree involves a two-year associate's degree followed by a bachelor's degree. This pathway allows you to begin your studies at a community college and then transfer to a four-year university to complete your bachelor's degree in your chosen engineering discipline. This can be a more affordable option, particularly if you are unsure of your specific engineering interests or wish to strengthen your academic record before entering a larger university program. Community colleges often provide smaller class sizes and more individualized attention, which can be beneficial as you begin your engineering studies. After completing the associate's degree, graduates typically transfer to a four-year institution, often with junior standing, to complete the remaining coursework required for their bachelor's degree.

Bachelor's Degree in a Another Field + Master's in Engineering

For those who already possess a bachelor's degree in a non-engineering field, a Master's degree in a specific engineering discipline offers an accelerated path to becoming an engineer. This option allows individuals with a strong background to transition into the field of engineering more quickly than starting a new four-year undergraduate degree. While this route can be efficient, it's important to note that some Master's programs may require prerequisite or "bridge" coursework to address any gaps between your existing background and fundamental engineering principles. These additional courses may ensure that you have the necessary foundation in areas like statics, dynamics, and materials science before beginning more specialized engineering topics. Notably, Systems engineering is often a pathway for those transitioning into engineering at the graduate level.

Professional Development

Even if you're not planning to become a professional engineer, cultivating an "engineering mindset" and developing core engineering skills can be incredibly valuable in any field. There are numerous avenues to explore, even without a formal engineering degree. Consider hands-on workshops and online courses that focus on specific skills like coding, design thinking, or problem-solving. Many organizations offer free or low-cost webinars and tutorials on technical topics, providing accessible learning opportunities. Look for "maker spaces" or community workshops where you can engage in hands-on projects and learn from others. Bootcamps, micro-certifications, and industry-recognized certifications can also provide focused training and demonstrate your competency in specific areas, boosting your resume and opening doors to new opportunities. Don't underestimate the power of self-directed learning; explore online resources, read books and articles, and follow experts in fields that interest you. By actively seeking out these diverse learning experiences, you can build a strong foundation in practical engineering skills and develop a knack for innovative problem-solving, regardless of your educational background.

Practical Experience

Lastly, practical experience is a powerful teacher, and it's essential for anyone looking to develop engineering-related skills, whether or not they're pursuing a formal degree. Look for opportunities to apply your knowledge in real-world scenarios. Even volunteer projects or community initiatives can offer valuable hands-on experience. If you're interested in a particular trade, explore apprenticeship programs or other on-the-job training opportunities. These programs allow you to learn from skilled professionals while gaining practical experience in a specific field. Shadowing engineers or technicians can also provide valuable insights into the day-to-day work involved in various engineering roles. Don't be afraid to reach out to local businesses or organizations to inquire about internship opportunities or other ways to gain practical experience. Building a portfolio of projects and experiences, even outside of a formal educational setting, can demonstrate your skills and commitment to potential employers or clients.

No matter which path you choose, imagine landing a rewarding career that allows you to use your creativity, problem-solving skills, and technical knowledge to make a real impact on the world.

Prompting Inquiry

Brain Teasers. These brain teasers are designed to sharpen your systems thinking, critical thinking and engineering-oriented skill. Try them and check your answers in Appendix D.

The Mysterious Message: You find a note with the following symbols: △□○◇. Each symbol represents a letter. △ = A. □ = T. ○ = O. What word does ◇ represent?

The Lost Kitten: A kitten is stuck at the top of a tall tree. Three people offer to help. A firefighter suggests a ladder. A construction worker suggests a crane. A child suggests something simpler. What does the child suggest?

The Empty Box: You have a box. It's empty. Can you make it full without putting anything inside it?

The Repeating Riddle: What is always coming, but never arrives?

The Mismatched Socks: You have a drawer full of socks. You have 10 white socks and 10 blue socks. It's dark, and you can't see the colors. How many socks do you need to take out to guarantee you have a matching pair?

The Broken Bike: Your friend's bike chain broke while you're miles from home. You have no tools. How do you get the bike moving again (even if it's not perfect)?

The Leaky Roof: A storm is coming, and there's a small leak in your roof. You don't have time for major repairs. What's a quick, temporary fix using common household items?

The Confusing Code: You find a coded message: 1-1-2-3-5-8-13-21. What does it mean?

The Island Escape: You're stranded on a deserted island with a box of matches. You need to build a signal fire, but all the wood you find is damp. How do you start the fire?

The Growing Problem: A patch of weeds in your yard doubles in size every week. If it takes 8 weeks to cover the entire yard, how long will it take to cover half the yard?

Chapter 5

Relationship Appeal

There's no way I could pay you back, but my plan is to show you that I understand. You are appreciated.

-Tupac Shakur

Key Topics

- Who are the people in your life?

- What value do you recognize in relationships, mentorships, teamwork, family, and friendships?

Let's take what may seem like a sight departure from the world of engineering to talk about something that's fundamental to everything we do: relationships. Yes, I'm serious! Even as I approach 40, I have realized that many adults (maybe even adults you know) misunderstand what a relationship really is. It's a word we use all the time, but do we truly get it?

Before we start, I want you to think about relationships for a second. What does the word "relationship" mean to you? Is it just about romance? Is it only about family? Or is there something more to it?

Relationships, in their broadest sense, encompass any connection you have with another person – from family and friends to classmates, teachers, mentors, and even those fleeting interactions with people you meet briefly. They're the fabric of our lives, the network of connections that shape our experiences and influence who we become. And while we often strive for the positive, supportive relationships that lift us up, it's important to recognize that relationships can also be challenging, even negative at times. Both the good and the not-so-good bring value, insights, and, perhaps most importantly, moments of self-discovery. The positive ones nurture our confidence and provide a sense of belonging, while the difficult ones can teach us resilience, boundaries, and what we truly value in our connections with others. Both types of relationships, in their own way, contribute to our growth and development, especially during the crucial years you are navigating right now.

Positive Relationships

Strong, positive relationships with mentors, family, and friends provide a crucial foundation of support and security for you. These relationships offer you a sense of belonging, acceptance, and unconditional love, creating a secure base from which you can confidently explore your identity, take healthy risks, and navigate the challenges of adolescence. Knowing you have a network of support empowers you to venture out and discover who you are.

This sense of belonging and support is further enhanced by the encouragement and positive feedback you receive from trusted individuals. Encouragement, validation, and positive feedback from trusted individuals play a significant role in boosting your self-esteem. Feeling valued, respected, and appreciated by those you care about helps build your self-confidence and self-worth. These positive interactions foster a healthy self-image and empower you to believe in your own abilities and potential.

Beyond boosting self-esteem, positive relationships also provide invaluable opportunities for skill development. Positive relationships provide opportunities for you to develop essential social and communication skills. Interactions with family, friends, mentors, and others allow you to practice effective communication, cultivate empathy, and learn constructive conflict resolution techniques. Through these interactions, you learn to navigate diverse social situations, express yourself clearly and respectfully, and understand different perspectives.

These skills are not only important in your personal life but also contribute to academic success. Positive relationships with teachers and mentors can significantly promote academic achievement. Supportive figures can provide guidance, celebrate academic milestones, and help you overcome learning challenges. These relationships can inspire you to strive for academic success, fostering a love of learning and a belief in your academic capabilities.

Moreover, the influence of positive relationships extends to your overall well-being and health. Positive role models within your network can encourage the adoption of healthy habits. Seeing mentors, family members, or friends prioritize healthy lifestyle choices—including balanced diet, regular exercise, and positive stress management—can motivate you to make similar choices. These positive influences can contribute to your overall well-being and long-term health.

To understand the power of positive relationships in your own life, consider this: Let's think about one positive relationship in your life. It could be with a family member, a close friend, a teacher, a coach, or anyone who makes you feel good about yourself. Take a moment to picture that person in your mind. What qualities do they have that you admire? What kinds of things do you do together? Now, consider how that connection makes you feel. Do you feel supported and encouraged? Do you feel like you can be yourself around them? Reflecting on the positive connections in your life can help you understand what you value in relationships and how those connections contribute to your overall well-being.

Negative Relationships

Negative relationships, characterized by bullying, criticism, or constant negativity, can significantly undermine your self-esteem. These experiences can inflict deep emotional wounds, creating feelings of insecurity, inadequacy, and self-doubt that can be difficult to overcome. The constant barrage of negativity can erode your sense of self-worth and make you question your abilities and value.

This erosion of self-worth can have a ripple effect on other aspects of your life, including your social interactions and choices. People involved in negative social circles may face increased pressure to engage in risky behaviors. Peer pressure within these groups can normalize or even glorify activities like substance abuse, delinquency, or unprotected sex. These behaviors can have devastating consequences for your physical and emotional well-being, leading to long-term health problems, legal issues, or emotional trauma.

Furthermore, toxic relationships can take a significant toll on your mental health. Toxic relationships can be a major source of stress and anxiety, sometimes even contributing to or exacerbating depression. Constantly feeling unsupported, criticized, manipulated, or even abused within a relationship can lead to feelings of isolation, hopelessness, and withdrawal. The emotional toll of these relationships can be overwhelming, impacting your overall mental health and well-being.

The impact of negative relationships can also extend to your academic performance. Negative relationships with teachers or peers can create a hostile and unproductive learning environment, hindering academic achievement.

People experiencing these challenges may disengage from schoolwork, skip classes altogether, or struggle to concentrate due to the emotional turmoil they are experiencing. The stress and anxiety caused by these relationships can make it difficult for you to focus on your studies and reach your full academic potential.

In an attempt to cope with the pain of negative relationships, some people may resort to unhealthy behaviors. People trapped in negative or abusive relationships may turn to unhealthy coping mechanisms as a way to manage the stress and emotional pain. These coping mechanisms, which can include substance abuse, self-harm, or disordered eating, can provide temporary relief but ultimately create more problems and further jeopardize your physical and mental health. These behaviors can become a dangerous cycle, making it even harder to escape the negative relationships and seek healthier support systems.

To better understand the impact of negative relationships, let's consider your own experiences. Think about a relationship in your life that's been challenging or even negative. It might be with someone at school, a family member, or even a friend where things have become strained. It's important to be honest with yourself here. What makes this relationship difficult? Are there frequent conflicts? Do you feel drained or unsupported after spending time with this person? Does this relationship make you feel bad about yourself in any way? Reflecting on these challenging relationships, while sometimes uncomfortable, can be incredibly valuable. It can help you understand your own boundaries, identify what you don't want in your connections with others, and ultimately learn how to navigate difficult interpersonal situations.

It's important to emphasize that not all relationships are black and white. Even positive relationships can have challenging moments, and negative ones might offer occasional support. However, the overall balance of a relationship is what matters most.

The Importance of Guidance

Parents, educators, and mentors play a vital role in guiding you through the complex world of relationships. By fostering open communication, offering consistent guidance, and providing unwavering support, adults can empower you to build healthy connections, recognize warning signs in unhealthy relationships, and develop the essential skills needed to create a strong and supportive network as you transition into adulthood. These relationships are particularly crucial during adolescence, a period of immense personal growth and self-discovery. The following sections will explore the unique value each type of relationship can bring to your life.

Mentorships

One particularly valuable relationship is mentorship. Mentors play a crucial role in your development, offering invaluable guidance and support as you navigate the often-turbulent waters of adolescence. They provide a safe space for you to discuss personal or academic struggles, explore career paths, and develop your sense of self. Beyond offering advice, mentors often serve as positive role models, inspiring you to set ambitious goals and strive for excellence. They can also expand your network, connecting you with new opportunities, people, and resources that can be instrumental in your growth and future success.

Teamwork

Beyond mentorship, teamwork experiences offer another crucial avenue for growth and development. Experiences in teamwork, whether through sports, clubs, or group projects, are essential for developing crucial life skills. Teamwork teaches you how to collaborate effectively, communicate clearly, compromise when necessary, and resolve conflicts constructively. These experiences also provide opportunities for you to discover and nurture your leadership potential, learning to motivate others, delegate tasks, and work towards a common goal. Furthermore, the challenges inherent in teamwork encourage creative problem-solving and critical thinking, as you learn to consider diverse perspectives and find solutions that benefit the entire team.

Family

Of course, family plays a foundational role in shaping who you become. Family provides the foundation upon which your sense of self is built. It often offers a bedrock of unconditional love, acceptance, and belonging, creating a safe and supportive environment where you feel understood and valued. Strong family relationships foster healthy communication skills, teaching you how to express yourself openly and respectfully, listen attentively, and navigate difficult conversations. Family dynamics also provide valuable lessons in conflict resolution, equipping you with the skills to manage disagreements constructively, compromise effectively, and find solutions that work for everyone involved.

Friends

Finally, friendships provide essential support and companionship during adolescence. Friendships are a vital source of social support, offering companionship, shared experiences, and a sense of belonging outside the family unit. Friends help you navigate the complexities of social situations, providing a sounding board for your thoughts and feelings. These relationships also play a key role in the development of your sense of identity, as you explore shared interests, discover who you are in relation to others, and build self-confidence. Friends serve as an important emotional outlet, offering a safe space for you to express your emotions, share anxieties, celebrate successes, and navigate the ups and downs of adolescence.

In summary, relationships, both positive and negative, are a powerful force in your life, especially now. They shape who you are, how you see the world, and the paths you choose to take. While we all crave those positive connections that lift us up and make us feel supported, even the challenging relationships can offer valuable lessons about ourselves and our boundaries. Navigating this complex landscape of human connection isn't always easy, and that's where the importance of guidance comes in. Whether it's the steady support of family, the wisdom of a mentor, the camaraderie of teamwork, or the honest feedback of a trusted friend, having someone to help you process your experiences, understand your emotions, and offer different perspectives can make all the difference. Remember, you're not alone on this journey. Lean on those who care about you, seek out guidance when you need it, and know that every relationship, in its own way, contributes to your growth and development.

Prompting Inquiry

Relationship Appeal. Explore what makes a relationship meaningful to you. Use the following chart (or create your own) to brainstorm the qualities you value in different types of relationships.

	Specific Examples		
Relationship	Quality 1	Quality 2	Quality 3
Mentorship			
Teamwork			
Family			
Friendship			

Gratitude Journaling. For some period of time, dedicate time each day for journaling about the positive relationships in your life. Each day, focus on one specific person you are grateful for and explore the reasons why. Go beyond simply stating their positive qualities; identify specific examples and experiences that highlight their impact on your life. Consider:

- What qualities do you admire most about this person?

- What specific memories or experiences demonstrate their positive influence on you?

- How does this person make you feel?

- What have you learned from this person?

- How has this relationship shaped who you are today?

- What are you most grateful for in this relationship?

Network Support Map. Brainstorm a list of people who provide support and encouragement to you. This could include family members, friends, teachers, coaches, mentors, or community members. In the center of the poster board, write "My Support Network." Draw lines radiating outward from the center. On each connection point (node), write the name of a person in your support network. Place the node on the lines radiating from the center, creating a visual representation of your network. Reflect on the map – e.g. the different types of support each person provides, the qualities you admire in your supporters, and how you can show your appreciation for them.

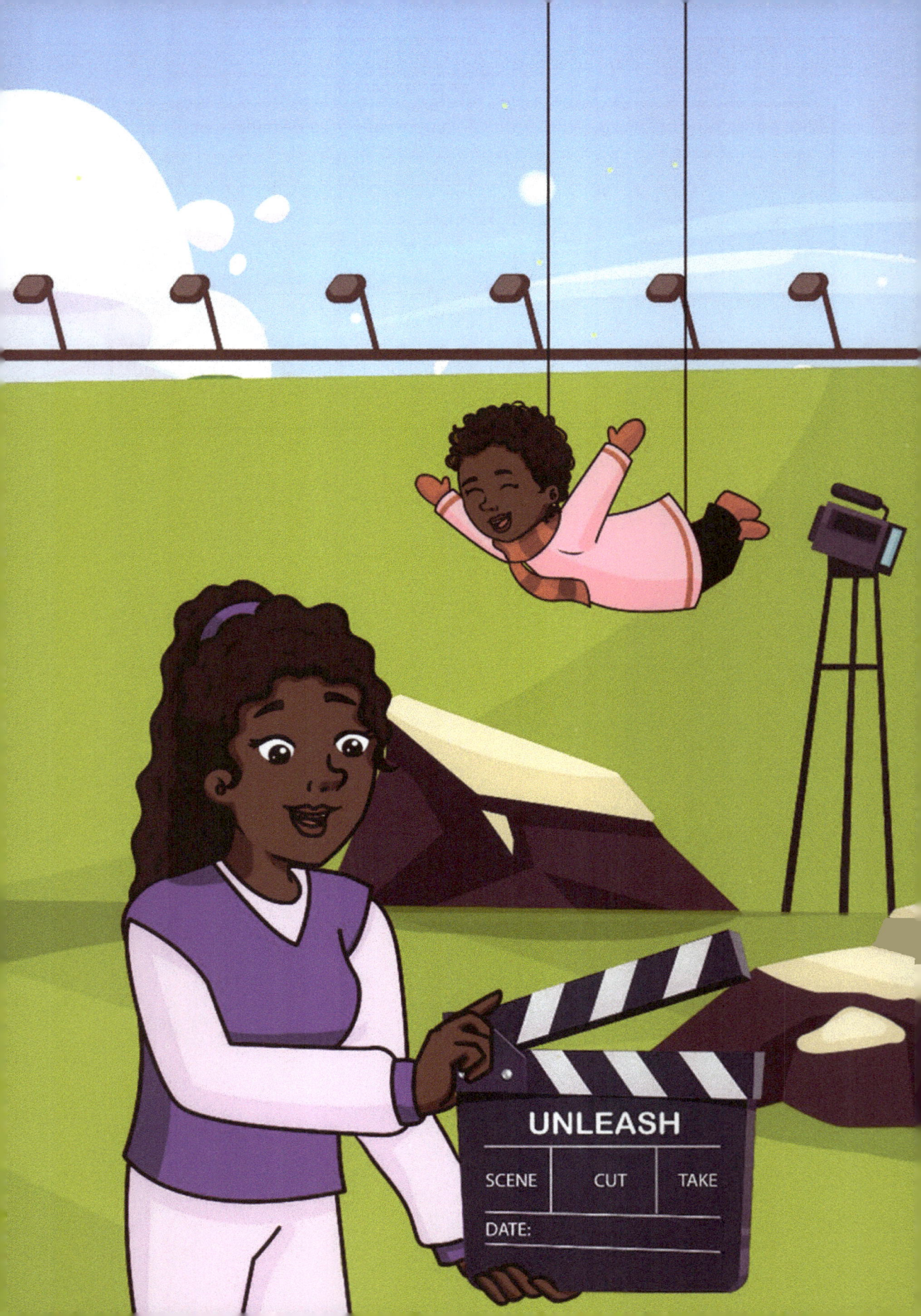

Chapter 6
Ready, Set, Action

Most people miss opportunity because it is dressed in overalls and looks like work.

-Thomas Edison

Key Topics

- What actions can you implement in the short-term?

- What age-appropriate opportunities for engagement are available?

The journey to becoming proficient in engineering systems is not always easy, but the ability to innovate and improve the world makes it undeniably worthwhile. But where do you start? It might seem like a long road ahead, but there are plenty of things you can do *now* to lay a solid foundation for your future. This chapter outlines practical steps you can take to explore your interests, develop essential skills, and gain valuable experience, setting you on the path to success. From academic preparation to extracurricular activities and real-world exploration, these actions will not only make you a stronger engineering applicant down the line but also help you discover what field is truly the right fit for you.

Some content may appear redundant, but this repetition is intentional, designed to reinforce the importance of taking action. Begin by seeking out introductory materials—look for articles, websites, or videos that explain fundamental engineering concepts in a clear and engaging way, making them accessible and easy to understand. Don't limit yourself to just one area; explore the diverse landscape of engineering disciplines, browsing through different fields to discover what sparks your curiosity and aligns with your interests. Connecting theory to practice is key, so look for real-world examples of how engineering principles are applied in everyday life, from reading about current projects and innovations to observing the technology around you. Never hesitate to ask questions—if you encounter something you don't understand, reach out to a teacher, mentor, or online community for clarification and support.

Building a Solid Foundation

Building a solid academic foundation is crucial, and we've already discussed the importance of challenging yourself with rigorous math and science courses when possible. If your school offers them and you are eligible, consider advanced placement (AP) classes or dual-enrollment programs at a local community college to gain a head start on college-level coursework. Regardless of the specific classes you take, prioritizing the development of strong study habits is paramount. Engineering programs are demanding, requiring dedication and effective learning strategies. Cultivate excellent time management skills to balance your academic workload and other commitments. Practice active learning techniques, such as summarizing material in your own words and teaching concepts to others, to deepen your understanding. Don't hesitate to seek assistance from teachers, tutors, or classmates when you encounter challenging material; building a support network is essential for academic success.

Develop Practical Skills

Developing practical skills is just as important as building a strong academic foundation. In today's engineering landscape, coding and programming proficiency is highly valuable across many disciplines. Learning a programming language like Python, Java, or C++ will not only enhance your problem-solving abilities but also expand your technical expertise. Computer-Aided Design (CAD) software is another essential tool in many engineering fields, so consider exploring introductory courses or online tutorials to familiarize yourself with the basics of CAD programs. Perhaps most importantly, seek out opportunities for hands-on learning whenever possible. Participating in robotics clubs, engaging in personal projects using LEGOs or other construction sets, and attending engineering-focused workshops are all excellent ways to solidify your understanding of engineering concepts and develop practical skills that you can apply to real-world challenges.

Explore Different Engineering Disciplines

Exploring the diverse landscape of engineering disciplines is crucial for finding the right fit for your interests and aptitudes. Numerous engineering fields exist beyond those covered in Chapter 3, each with its own unique focus, challenges, and opportunities. Take the time to research various specializations. Investigate what each field entails, the types of problems engineers in that discipline tackle, and the skills and knowledge required for success. This research will help you narrow down your options and identify the area of engineering that best aligns with your passions and career aspirations.

Here are 100 diverse engineering disciplines, spanning traditional fields to emerging specializations.

Core & Traditional Disciplines

1. Aerospace Engineering
2. Agricultural Engineering
3. Architectural Engineering
4. Automotive Engineering
5. Biomedical Engineering
6. Bioengineering
7. Chemical Engineering
8. Civil Engineering
9. Computer Engineering
10. Construction Engineering
11. Electrical Engineering
12. Environmental Engineering
13. Fire Protection Engineering
14. Geological Engineering
15. Industrial Engineering
16. Manufacturing Engineering
17. Marine Engineering
18. Materials Science and Engineering
19. Mechanical Engineering
20. Mechatronics Engineering
21. Mining Engineering
22. Nuclear Engineering
23. Ocean Engineering
24. Petroleum Engineering
25. Robotics Engineering
26. Software Engineering
27. Structural Engineering
28. Systems Engineering
29. Telecommunications Engineering
30. Transportation Engineering

Interdisciplinary & Cross-Disciplinary Fields (Often Combinations)

31. Bioinformatics
32. Computational Engineering
33. Data Engineering
34. Environmental Health Engineering
35. Human-Computer Interaction Engineering
36. Information Engineering
37. Knowledge Engineering
38. Materials Characterization Engineering
39. Neuroengineering
40. Pharmaceutical Engineering
41. Rehabilitation Engineering
42. Smart Grid Engineering
43. Supply Chain Engineering
44. Systems Integration Engineering
45. Computational Materials Engineering

Specialized & Emerging Disciplines

46. Acoustical Engineering
47. Aerospace Systems Engineering
48. Agricultural & Biosystems Engineering
49. Applied Engineering
50. Architectural Acoustics Engineering
51. Automation Engineering
52. Biochemical Engineering
53. Biological Engineering
54. Building Services Engineering
55. Ceramic Engineering
56. Coastal Engineering
57. Computer Systems Engineering
58. Control Systems Engineering
59. Cryogenic Engineering
60. Design Engineering
61. Earthquake Engineering
62. Ecological Engineering
63. Energy Engineering
64. Engineering Management
65. Engineering Physics
66. Facilities Engineering
67. Food Engineering
68. Forensic Engineering
69. Foundation Engineering
70. Genetic Engineering
71. Geotechnical Engineering
72. Human Factors Engineering (Ergonomics)
73. Hydraulic Engineering
74. Instrumentation and Control Engineering
75. Irrigation Engineering
76. Laser Engineering
77. Logistics Engineering
78. Manufacturing Systems Engineering
79. Naval Architecture and Marine Engineering
80. Network Engineering
81. Nanotechnology Engineering
82. Occupational Safety and Health Engineering
83. Optical Engineering/Photonics
84. Packaging Engineering
85. Plastics Engineering
86. Power Engineering
87. Process Engineering
88. Product Design Engineering
89. Reliability Engineering
90. Safety Engineering

91. Satellite Engineering

92. Software Systems Engineering

93. Solar Energy Engineering

94. Sports Engineering

95. Sustainable Engineering

96. Textile Engineering

97. Thermal Engineering

98. Transportation Planning and Engineering

99. Urban Planning and Engineering

100. Water Resources Engineering

Read Engineering Magazines and Articles

With such a diverse range of engineering disciplines, staying current with the rapid pace of innovation is essential, and reading engineering magazines, articles, and reputable websites is a great way to do this. These resources offer insights into the latest advancements, emerging technologies, and cutting-edge research across various engineering disciplines. Regularly engaging with this content will not only broaden your overall engineering knowledge but also expose you to specific areas that might spark your interest and inspire you to explore further. Actively seeking out and digesting this information will keep you informed and help you identify potential career paths or specializations that align with your passions.

To get started, I've listed several key resources. Since online links can change frequently, I've provided detailed descriptions instead of direct links. This will allow you to easily find the most up-to-date versions using a search engine or library database.

Magazines & Print Publications

- **Engineering.com:** More than just a magazine, Engineering.com is a comprehensive online resource hub. It offers articles, news, white papers, and community forums covering various engineering disciplines, making it an excellent starting point for exploring the breadth of the field and connecting with others. It's a valuable resource for staying up-to-date on industry trends and emerging technologies.

- **Design World:** Design World magazine focuses specifically on mechanical and design engineering. It showcases innovative products, cutting-edge technologies, and practical design solutions. It's a great resource for inspiration and insights into the latest advancements in product design and manufacturing.

- **IEEE Spectrum:** Published by the prestigious Institute of Electrical and Electronics Engineers (IEEE), IEEE Spectrum offers in-depth coverage of a wide range of electrical and computer engineering topics. It provides expert analysis of current trends, emerging technologies, and the impact of these technologies on society. It's a valuable resource for both practicing engineers and those interested in learning more about the field.

- **Popular Mechanics:** This is a classic magazine known for its accessible approach to science, technology, and engineering. Popular Mechanics explains complex concepts in a way that's easy to understand. It often features articles on DIY projects, automotive technology, and the latest gadgets, making it a great choice for anyone interested in the practical applications of engineering.

- **E+T Magazine (Engineering & Technology):** Published by the Institution of Engineering and Technology (IET), E+T Magazine provides a global perspective on engineering, featuring articles on various engineering fields. It covers emerging technologies, industry news, and the role of engineering in addressing global challenges, making it a valuable resource for understanding the broader context of engineering practice.

Articles & Online Resources

- **"10 Amazing Engineering Feats of the 21st Century" (Various Sources):** Search for articles with the title "10 Amazing Engineering Feats of the 21st Century" (or similar variations) to discover remarkable examples of recent engineering achievements. These articles often highlight the ingenuity, innovation, and problem-solving skills involved in tackling complex challenges, showcasing the real-world impact of engineering. Be sure to compare different articles on this topic to gain a broader understanding of the diverse projects and perspectives within the field.

- **Khan Academy, Engineering Section:** This resource offers introductory videos and clear explanations of core engineering concepts, making complex ideas more accessible. This is a valuable resource for those exploring engineering or want to reinforce their understanding of basic principles.

- **Explain that Stuff:** Explain that Stuff is a website dedicated to providing clear and concise explanations of various engineering and scientific concepts. It breaks down complex topics into easy-to-understand language, making it a valuable resource for anyone curious about how things work. Whether you're interested in learning about bridges, electricity, or the internet, this resource can help in grasping the underlying engineering principles.

- **Science Buddies:** Science Buddies offers a wide range of hands-on science and engineering project ideas, from simple experiments to more advanced engineering challenges, encouraging practical learning and exploration. Science Buddies also provides resources and guidance to help individuals through the scientific method and engineering design process, fostering critical thinking and problem-solving skills.

- **TryEngineering (IEEE):** TryEngineering is an initiative of the Institute of Electrical and Electronics Engineers (IEEE) that offers a wealth of materials, including engaging activities and detailed information on various engineering fields to inspire the next generation of engineers.

- **National Academy of Engineering (NAE):** The NAE is a prestigious organization that recognizes outstanding engineers and provides leadership in engineering education, research, and practice. The NAE website offers insights into grand challenges facing society, highlights groundbreaking engineering achievements, and provides resources on engineering careers and education. It's a valuable resource for anyone interested in learning about the impact of engineering on society and exploring the future of the profession.

Remember, this is just a starting point. There are many other great resources available online, in print and within your community. The key is to be curious, explore different options, and find the resources that best suit your interests and learning style.

Gain Real-World Experience

Gaining real-world experience is invaluable for anyone aspiring to better understand engineering and systems. Summer internships offer a fantastic opportunity to immerse yourself in the professional world, applying your classroom knowledge to practical settings and gaining exposure to specific engineering fields that pique your interest. Beyond formal internships, volunteering for engineering-related projects in your community can also provide valuable hands-on experience. This could involve assisting with science fairs, contributing to building projects for local organizations, or participating in environmental clean-up efforts with an engineering focus. Finally, shadowing an engineer in your area of interest, if possible, can give you a first-hand glimpse into their daily tasks and responsibilities, offering valuable insights into the realities of the engineering profession and helping you solidify your career aspirations.

Develop Soft Skills

Developing strong soft skills is just as critical as mastering technical knowledge for aspiring engineers. Effective communication is essential, as engineers must be able to articulate complex ideas clearly and concisely, both verbally and in writing. Participating in activities like debate clubs, writing programs, or public speaking opportunities can significantly enhance these communication skills. Furthermore, teamwork and collaboration are paramount in the field of engineering. Engineering projects are rarely solitary endeavors, requiring individuals to work effectively with others, delegate tasks efficiently, and resolve conflicts constructively. Developing these teamwork skills through team sports, group projects in school, volunteer activities, or any collaborative effort toward a shared goal will be invaluable in your future engineering career.

Above all, don't be afraid to show your curiosity and passion for the field. Express your enthusiasm for engineering and systems whenever you have the opportunity. Talk to your teachers, counselors, and mentors about your career goals and interests. Genuine passion is contagious and can open doors you never imagined. Whatever it is that you choose to do, it is my firm belief that you should do so wholeheartedly. When you approach your pursuits with genuine enthusiasm and dedication, you not only increase your chances of success but also inspire those around you. Let your passion fuel your journey and guide your path toward a fulfilling future.

Prompting Inquiry

Engineering Disciplines. Select 5 disciplines from this chapter's list and research each one, considering:

- What kinds of problems do engineers in this field solve?

- What skills and knowledge are crucial, and do these align with your strengths and interests?

- Where can you study this discipline? (Find at least one college or university that offer the degree program.)

Passion Portfolio. Explore your interests. Begin by creating a "Passion Inventory" – list activities you enjoy, things you are curious about, and skills you possess. Reflect on past experiences where you felt engaged and fulfilled. Identify recurring themes in these interests.

- Using this process, choose 3 personal passions or interests that truly resonate with you.

- Create a "mini-project" for each of your passions aiming to demonstrate how you could use them in your career (e.g., a short story for writing, a song for music, a photo series for photography, a simple website for coding, a dance routine for dance, a styled look for fashion).

- Document your journey in a reflective journal or sketchbook (paper or digital), noting personal insights and connections to your values.

- Share your portfolio with a parent, mentor or friend.

- As you explore these diverse passions, consider: Do you see any connections between your passions and the engineering disciplines listed in this chapter?

Community Search. Visit the websites of at least one college or university that interest you (or are geographically accessible). Also consider programs offered by other organizations in your local community. Look specifically for programs geared towards your age group, such as:

- Summer programs

- Pre-college programs

- STEM/STEAM camps

- Dual enrollment/early college programs

- Engineering-focused clubs or workshops

For each program you find, investigate the following:

- What is it all about? Briefly describe the program's focus and what participants will learn or do.

- Who can apply? Are there age or grade level restrictions? Are there specific academic requirements (e.g., GPA, coursework)?

- When are applications due? Mark the deadlines on your calendar!

- Who's the go-to person? Find the contact information (email address or phone number) for the program coordinator or admissions contact.

- What do you need to apply? List all the required application materials (e.g., application form, transcripts, essays, letters of recommendation).

Note: Even if you're not 100% committed to engineering or if you have concerns about resources, I strongly encourage you to apply to at least one program that sparks your interest. And here's why- Participating in a program doesn't mean you have to become an engineer. It's a chance to explore the field and see if it's a good fit for you. You might be surprised by the scholarships, financial aid, or other resources available to support you. Many programs are designed to be accessible to participants from all backgrounds.

In addition, participating in pre-college programs looks great on college applications and can demonstrate your initiative and interests in general. You'll also meet other people who may share your interests and build connections with faculty and professionals in the field.

Chapter 7
Unleash the Best of You

It matters not how strait the gate,

How charged with punishments the scroll I am the master of my fate,

I am the captain of my soul.

-William Ernest Henley

Key Topics

- Have you ever hesitated to pursue something because it seemed out of reach, unpopular, or you doubted your own capabilities?

- How can you believe in your potential to become an engineer or whatever you choose to become?

Let's begin this chapter with a frank acknowledgment: we've all been there. That moment when a dream feels so distant, so improbable, that we hesitate to even reach for it. Maybe it's because the goal seems too ambitious, the path too challenging. Or perhaps it's the fear of judgment, the worry that others will see your aspirations as unrealistic or even "uncool."

Before we get too far into this chapter, take a moment to describe a specific instance where you believe you missed an opportunity to learn something new. What circumstances led to this, and what do you think you could have done differently?

```



```

Similarly, describe a specific situation where you recognize you missed an opportunity to challenge yourself. What were the circumstances, and what do you think prevented you from taking on the challenge?

```



```

Doubt can creep in from many sources, sometimes even from well-meaning individuals who question your abilities or the validity of your dreams. And how often have you heard stories of success that magically materialized without effort? The truth is, achieving anything worthwhile requires belief—belief in yourself, belief in your potential, and belief that you are capable of more than you might think. This chapter explores those feelings of doubt and hesitation, particularly as they relate to the pursuit of engineering systems dreams. We will dive into how to overcome these internal and external obstacles, challenge limiting beliefs, and cultivate the confidence needed to pursue your engineering aspirations, even when the path ahead seems daunting.

How many times have you hesitated to pursue something that sparked your interest, simply because it felt too far-fetched or unattainable? Perhaps it was a dream of becoming a musician, a scientist, or even an engineer. That nagging voice of doubt whispers that it's "not for you," that you "don't have what it takes," or that the path is too difficult. Or maybe the fear of social judgment holds you back. The worry that your aspirations will be seen as unpopular, uncool, or just plain weird can be a powerful deterrent, especially during adolescence when fitting in often feels paramount. But consider this: what if the very thing that seems out of reach is actually within your grasp? What if the "unrealistic" dream is the one that's meant to be yours? And what if embracing the "unpopular" path is the key to unlocking your true potential and finding genuine fulfillment?

Doubt, unfortunately, can sometimes come from external sources. Have you ever shared a dream, a goal, or an aspiration only to be met with skepticism, discouragement, or even outright disbelief? Perhaps a well-meaning family member, a friend, or even a teacher has questioned your capabilities, suggesting that your chosen path is too challenging, too competitive, or simply not a good fit for you. These moments can be particularly disheartening, especially when the doubt comes from someone you respect or trust. It's easy to internalize these messages, to let their doubts become your own. But it's important to remember that other people's perceptions are not necessarily a reflection of your true potential. Their limitations are not yours. Their fears are not yours. Your dreams are yours, and you have the power to pursue them, regardless of the naysayers.

But beware. How often have you heard stories of "success" that magically materialized without effort, struggle, or setbacks? Probably never. The narratives we often consume, especially in popular culture, can sometimes paint a distorted picture of achievement, making it seem as though success is a sudden and effortless arrival rather than the culmination of dedication, perseverance, and often, a healthy dose of failure. We see the finished product, the polished performance, the award-winning innovation, but rarely do we witness the countless hours of practice, the challenges overcome, the rejections faced, the failed system tests, and the lessons learned along the way. The truth is, genuine success in any field, including engineering, is rarely a stroke of luck. It's built upon a foundation of hard work, resilience, and a willingness to learn and grow, even when the path gets tough.

So, how can you believe that you can master an engineering systems mindset? It starts with recognizing that engineering isn't some exclusive club reserved for a select few. It's a diverse and dynamic field open to anyone with curiosity, a passion for problem-solving, and a willingness to learn. Believe in your capacity to develop the necessary skills and knowledge. Embrace the idea that challenges are opportunities for growth, and that setbacks are simply stepping stones on the path to success. Surround yourself with supportive individuals who believe in your potential, and seek out mentors who can offer guidance and encouragement. Most importantly, believe in yourself. Believe in your ability to learn, to adapt, and to persevere. Believe that your unique perspective and talents are valuable assets to the engineering world. The journey may not always be easy, but with dedication and self-belief, you can achieve your engineering aspirations.

I invite you to consult the engineering systems creed in Appendix B. Downloading a copy of this creed will serve as a constant reminder of the dedication and principles required to excel in engineering systems.

Believe in Yourself

Your belief in your ability to succeed—your self-efficacy—is a key factor in pursuing any goal, especially in a demanding field like engineering. A strong sense of self-efficacy can be a powerful motivator on your journey.

High self-efficacy offers numerous benefits for those aspiring to embrace engineering systems. It fuels motivation and persistence, driving you to embrace challenging math and science courses, actively participate in engineering-related activities, and persevere through difficult problems. This stems from a core belief in your ability to learn and master the skills essential for engineering success. Furthermore, high self-efficacy can significantly reduce anxiety and fear of failure. If you have a strong sense of self-efficacy, you are less likely to be discouraged by setbacks, viewing failures not as defeats but as valuable opportunities for learning and improvement. This positive mindset allows you to approach challenges with greater confidence and resilience.

This confidence and resilience translate into proactive behavior and goal setting. Beyond motivation and resilience, high self-efficacy empowers you to set ambitious goals and proactively develop plans to achieve them. You are more likely to seek out learning opportunities and challenges that will further enhance your skills and knowledge. This proactive approach to learning, coupled with a belief in your capabilities, often translates to improved academic performance. Studies consistently demonstrate a strong correlation between self-efficacy and academic achievement, showing that those with high self-efficacy in engineering are more likely to excel in math, science, and other related subjects, setting themselves up for success in their chosen field.

Conversely, a lack of self-efficacy can create significant barriers to pursuing an engineering career. Low self-efficacy can significantly hinder you. If you doubt your abilities in math, science, or related areas, you may avoid challenging courses and shy away from engineering-related activities altogether, fearing failure. This self-doubt can create a cycle of limited exploration, preventing you from discovering the diverse fields within engineering and potentially closing doors to a viable and fulfilling career path simply due to perceived limitations. You may not even consider engineering as an option because you lack the confidence in your own skills.

This self-doubt can also impact your ability to persevere when faced with challenges. When faced with setbacks, you may be more likely to give up easily, believing you lack the capability to master new skills or solve complex problems. This lack of resilience can be a major impediment to success in engineering, a field that often requires persistence, problem-solving skills, and a willingness to learn from mistakes. Ultimately, low self-efficacy can create a self-fulfilling prophecy, preventing you from reaching your full potential and achieving your aspirations.

Keys to Building Self-Efficacy

- Embrace Mastery Experiences: Seek out opportunities to experience success, starting with challenges you know you can conquer, and gradually increasing the difficulty as your skills grow. Celebrate your accomplishments, no matter how small, and track your progress, recognizing how far you've come. This helps you build confidence in your abilities and motivates you to keep learning. Understand that even "failures" are valuable learning experiences.

- Welcome Positive Reinforcement: Appreciate encouragement and praise for your effort, your perseverance, and your problem-solving skills, not just for perfect results. Understand that it's the learning process that truly matters, and strive to maintain a growth mindset, always believing you can improve.

Offer positive reinforcement to others, recognizing their efforts and celebrating their growth.

- Connect with Role Models and Mentors: Actively seek out connections with engineers and professionals in fields that interest you. Seeing successful people who share your background or experiences is incredibly motivating. It shows you that a career in engineering, or any field you choose, is attainable. Learn from their journeys, ask them questions, and value their guidance.

- Explore Engineering Concepts: Jump at opportunities to participate in robotics clubs, engineering summer camps, workshops, or even online challenges. These hands-on experiences spark your interest, develop your skills, and show you the practical applications of engineering principles. Look for ways to incorporate engineering systems thinking into your everyday life, from tinkering with gadgets to solving puzzles.

- Focus on Your Strengths: Take time to identify your talents and strengths, whether they're in math, science, creativity, problem-solving, teamwork, or something else entirely. Explore how these strengths can be applied to engineering and other fields you're interested in. Knowing what you're good at gives you a solid foundation to build upon.

- Set Realistic Goals: Work with mentors, teachers, or even family members to set achievable goals related to your aspirations. These goals might involve improving your grades in specific subjects, participating in a competition, learning a new coding language, or building a personal engineering project. Breaking down big dreams into smaller, manageable steps makes them feel less daunting and keeps you motivated. Celebrate each milestone you reach, no matter how small, as a step closer to your bigger goals.

Now that you've explored several approaches for building self-efficacy, think about your specific strengths and challenges. Which do you anticipate being easiest and hardest to apply in your own life? What factors influence your perspective?

Can you think of additional ways you can build self-efficacy in general, and for engineering systems specifically? Why do you think these additional strategies would be effective?

From engineering mindsets to understanding relationships, we've covered a lot of ground in this book so far. But the most important takeaway is this: you have the power to shape your own destiny. Self-efficacy is the foundation upon which you'll build your future, whatever path you choose. It's about believing in your ability to learn, grow, and make a difference. So, take what you've learned, embrace your potential, and go out there and create the life you envision. You've got this!

Prompting Inquiry

Everyday Problems. Observe your surroundings and identify a problem you or others experience regularly. This could be related to organization, efficiency, accessibility, or any other area of daily life. For example, you might consider the lack of convenient storage in your room, the difficulty of carrying multiple bags, or the challenges of navigating a crowded space.

-   **Define the Problem:** Clearly describe the problem, including who it affects, how often it occurs, and what its impact is.

-   **Brainstorm Solutions:** Generate at least three distinct solutions to the problem. Don't be afraid to think outside the box!

-   **Sketch and Visualize:** Create sketches or diagrams to illustrate your proposed solutions. Consider different perspectives and details.

-   **Evaluate and Refine:** Analyze each solution, considering its functionality, aesthetics, user needs, cost, and feasibility. Which solution do you believe is the most effective and practical? Why?

-   **Present Your Solution:** Briefly describe your chosen solution, explaining its key features and benefits. If you have the opportunity, try presenting your solution to others. This can be a great way to get valuable feedback and make your design even better.

Vision Board (Digital/Physical). Create a vision board that reflects your belief in yourself and your positive outlook on the future. This board should be a visual representation of your dreams, goals, and aspirations. Consider all aspects of your life – personal growth, relationships, career, education, creativity, health, and well-being. Use the things we've discussed in this book as needed but do not feel constrained by it. It's your vision for your own life.

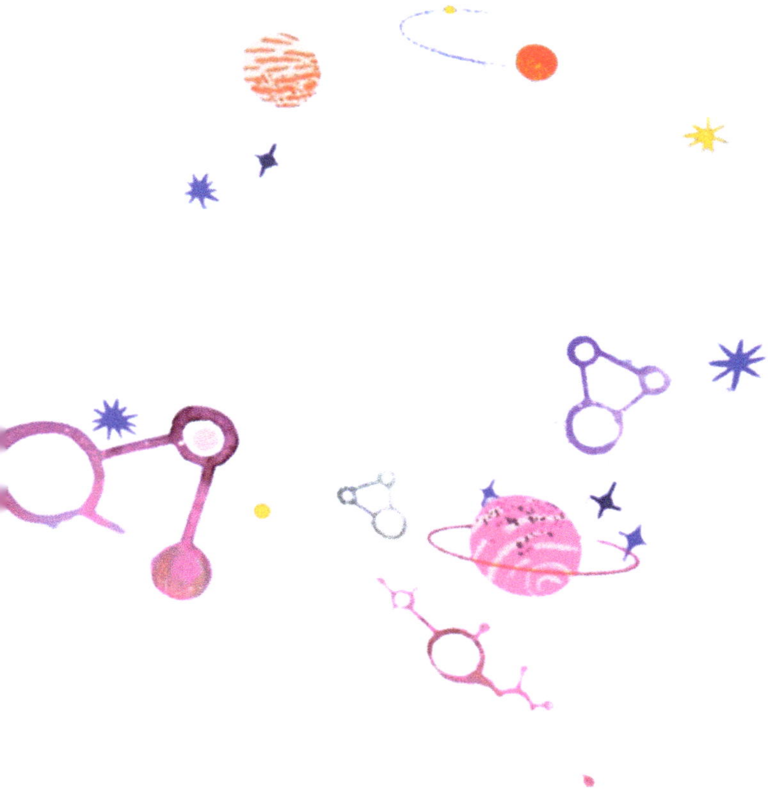

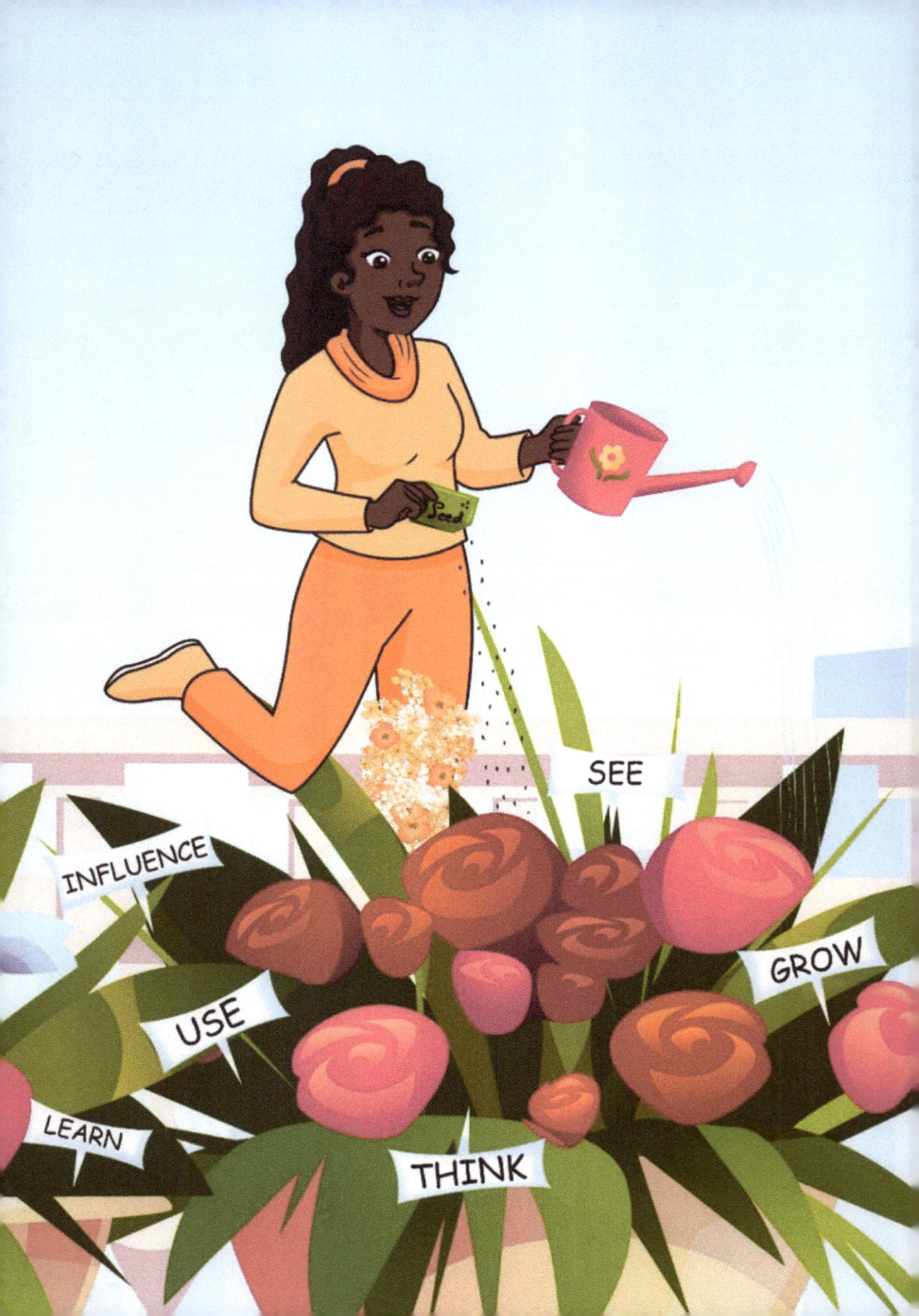

Chapter 8

And You're Off...

I have begun everything with the idea that I could succeed, and I never had much patience with the multitudes of people who are always ready to explain why one cannot succeed.

-Booker T. Washington

Key Topics

- How can self-awareness shape your path?

- Why is cultivating future growth essential for your success?

You don't know what you don't know.

It's a simple phrase, but it holds a powerful truth. Think about it: your world, right now, is shaped by your experiences, the people you know, and the things you've been exposed to. But that's just a sliver of what's out there. Your parents, your family, even your closest friends – they have their own limited perspectives too. There are entire worlds of ideas, cultures, and possibilities beyond what you've experienced so far. Seeking awareness isn't about saying your current world is bad; it's about recognizing that it's just one world among many. It's about opening your eyes, asking questions, and venturing beyond your comfort zone. Don't let your current reality define the limits of what you can achieve or who you can become. There's so much more to discover, and seeking awareness is the first step to unlocking it all.

This book has been an invitation—an invitation to see yourself not just as you are today, but as the architect of your own future. It's been an attempt to show you that the possibilities are truly limitless, that you can do anything you set your mind to. While many of the ideas and examples have been explored through the lens of engineering, the real goal has been much bigger than simply steering you toward a specific career path. It's been about igniting within you the power of an engineering systems mindset and the invaluable skills of systems thinking. These two things, working in tandem, are like a superpower. They're the tools you'll need to navigate the complexities of the world and propel yourself forward in any pursuit you choose, whether it's engineering, art, business, activism, or something entirely new.

Because here's the undeniable truth: systems are everywhere. They're not just in machines or technology; they're in our communities, our governments, our economies, even in the way we interact with each other. From the intricate workings of a cell to the global flow of information, everything is connected, nearly everything is a system. Understanding how these systems work, how the parts influence the whole, is crucial for success in virtually any field. By equipping yourself with this understanding, by learning to analyze, optimize, and innovate within systems, you're already positioning yourself for better outcomes, no matter what path you take. You'll be able to see connections others miss, anticipate challenges before they arise, and develop solutions that are both effective and sustainable. This is the power of a systems-thinking mindset, and it's a power you now possess.

Sprout Where you are Planted

My entire career, my entire enterprise, has been built on a simple but powerful idea: knowing where you stand, and then moving forward. It's the engine of progress, the key to unlocking potential. And it's a concept I want to add to your toolkit before we part ways, because it's just that important. What does it mean, "knowing where you stand?" For me, as a Black woman navigating a world that wasn't always designed with me in mind, it meant acknowledging the impact of history, systemic inequalities, and community influences on my own identity, my perceived capabilities, and even my own behavior. It meant confronting the realities of the past and present, the challenges and the biases, and understanding how they shaped my starting point. That was where I stood. But here's the crucial part: accepting that reality wasn't about accepting defeat. It was about embracing that person, the whole person, with all her strengths and her weaknesses, her resilience and her vulnerabilities. It was about recognizing the power within me to transcend limitations. That self-awareness, that honest appraisal of where I stood, provided the unshakeable foundation I needed to move forward – to plan strategically, to accomplish ambitious goals, to achieve success on my own terms.

Throughout this book, I've tried to guide you on a similar journey of self-discovery. I've hoped to help you discern and understand your positioning in the world, to recognize your own unique strengths and the challenges you might face. Because knowing where you stand – understanding your values, your passions, your potential, and the forces that might try to hold you back – is the first step toward true empowerment. It's the launching pad for your own journey forward. I truly believe you've got this! I know you've got this, deep down, and soon, very soon, the world will see it too. And that "it" – that power, that potential, that unique brilliance – will shine through no matter what path you choose. Whether you become an engineer, an artist, an entrepreneur, or something entirely your own, that self-knowledge, combined with the tools and mindsets we've explored, will be your greatest asset.

Just remember these guiding principles as you embark on your journey:

- Never stop asking "why" and "how." Be a relentless explorer, taking things apart (carefully, of course!) to understand their inner workings and putting them back together, sometimes even creating something new. Embrace the joy of discovery and the thrill of figuring things out.

- Don't shy away from the tough stuff. Difficult tasks are not roadblocks; they're opportunities to learn, grow, and prove what you're capable of. They are invitations to expand your limits and redefine what's possible. Remember that even the most brilliant minds faced setbacks and learned from their mistakes. Resilience is key.

- There's rarely just one "right" answer. Let your imagination soar, explore different possibilities, and dare to think outside the box. Creative solutions often come from unexpected places.

- Engineering, and life in general, isn't a solo act. Learn to communicate your ideas clearly and effectively, listen actively to others, and work collaboratively as part of a team. Diverse perspectives lead to stronger solutions.

- The world is constantly changing, and engineering is no exception. Cultivate a growth mindset, believing that your abilities and knowledge can always be expanded. Never stop learning, exploring, and adapting to new challenges and opportunities.

The world desperately needs your unique ideas, your boundless creativity, and your problem-solving prowess. You have the potential to make a real difference, to shape the future in ways you can't even imagine yet. So, go forth, wherever you are planted, and let your engineering systems spirit blossom. Nurture that systems-thinking mindset, because it's a powerful tool that will serve you well in any path you choose. Explore your passions, embrace the unknown, and use the skills and knowledge you've gained to build the future you envision. The world is waiting to see what you'll create.

Personal Challenge. Choose a personal challenge that is meaningful to you, achievable yet stretches you slightly, and within your control. It could be anything from learning a new skill (coding, playing an instrument, public speaking), improving a physical ability (running a 5k, mastering a yoga pose), tackling a personal goal (organizing your room, reading a certain number of books), or contributing to your community (volunteering at a local shelter, starting a small fundraising campaign). The key is that you choose it and are invested in it.

Then step through the following steps-

- **Planning:** Break down the challenge into smaller, manageable steps. Instead of "learn to play guitar," it's "learn three basic chords this week," then "learn a simple song next week," etc. This makes the overall goal less daunting and allows for regular wins.

- **Timeline:** Formulate a practical timeline that maps out your challenge, specifying a final goal and individual goals for each stage of the process. **Recall your ASTS, some activities can be completed in parallel, so consider which tasks can be done simultaneously to maximize efficiency.** This helps with time management and provides a sense of structure.

- **Action:** Work on your challenge, following your plan. You are encouraged to document your progress, whether through a journal, a blog, videos, or even just notes on your phone. This provides tangible evidence of your efforts.

- **Support:** Seek guidance and encouragement as needed, but avoid letting someone else do the work for you. The focus is on your own accomplishments. You may seek help troubleshooting if you encounter obstacles, but feel empowered to find solutions.

- **Adaptation:** Life happens! If your plan needs adjusting, that's okay. The ability to adapt and overcome setbacks is a vital part of building self-efficacy. At any time, re-evaluate your plan, if needed, and make necessary changes.

-Reflection:

- **Successes:** When you achieve a milestone or complete the challenge, celebrate your success! Reflect on what you did well, how you overcame challenges, and what you learned about yourself.

- **Challenges:** Even if you don't fully achieve your initial goal, there's still value in the experience. Reflect on what you learned from the process, what you could do differently next time, and what strengths you discovered. Understanding "failures" as learning opportunities is essential for building resilience and sustained self-efficacy.

- **Future Goals:** You are encouraged to think about how you can apply the skills and mindsets leveraged during this challenge to other areas of your life.

Appendices

Appendix A. Essential Terminology for Engineering Systems

Abstract System	Feedback	Prediction
Abstraction	Flow	Product
Algorithm	Formal	Programming
Alternative	Model	Property
Ambiguity	Function	Prospection
ASTS	Heuristics	Reductionism
Autonomy	Homogeny	Risk
Behavior	Human Factors	Root Cause
Boundary	Hypothetical	Service
Complexity	Ideology	Simplification
Component	Induction	Social
Conceptual Model	Inferential	Stakeholder
Constraint	Input	Stock
Creativity	Interconnection	Structure
Deduction	Intrarelationships	Subsystem
Delay	Lifecycle	Sustainment
Dependency	Linear	Symptom
Discipline	Mental Model	System of Systems
Economics	Metaverse	System
Efficiency	Natural	Systems Engineering
Element	System	Technical
Emergent	Optimize	Technology
Engineered System	Output	Tolerance
Engineering	Paradox	User
Enterprise	Perspective	Validate
Environment	Political	

Appendix B. Unleashed Engineering Systems Power Creed

Unleashed- The Creed

I have unleashed my engineering systems power.

I am a learner, a creator, and a problem-solver that believes in my potential.

My mind is a powerful tool, capable of understanding complex ideas and finding innovative solutions.

I am determined to know where I stand, so that I can move forward strategically and intentionally.

I embrace challenges not as roadblocks, but as opportunities to grow stronger and wiser.

I am not afraid to ask questions, to experiment, or to sometimes fail, because I know that every experience, good or bad, teaches me something valuable.

I take time to reflect on my experiences, analyze what happened, and gain insights that help me improve.

I choose to focus on the positive, training my mind to spotlight the good in situations, knowing that even in setbacks, there are lessons to be learned.

I see the world not as a collection of isolated parts, but as a system of interconnected pieces.

I see the big picture and understand how the parts fit together.

I understand that everything is related, and that even small actions can have a big impact.

I analyze situations critically, creatively, and strategically to identify root causes, and design effective solutions.

I am open to new and different ideas, eager to explore new experiences, perspectives, and knowledge.

I engage with diverse communities, understanding that learning from people with different backgrounds enriches my own understanding.

I am capable of achieving anything I set my mind to, including becoming an engineer, a scientist, an artist, or anything else I dream of.

There are no limits to what I can learn or what I can create. I will not let self- doubt hold me back.

I will embrace my unique talents and use them to make a positive difference in the world.

I am resilient, resourceful, and I wholeheartedly believe in myself.

I am the architect of my own future, and I choose to build it with confidence, courage, and a systems-thinking mindset.

Appendix C. Areas of Systems Thinking Skills (ASTS) Development Strategies

ASTS. Multiple Perspectives

- Listen Up: Really listen when other people are talking, even if you think you already know what they're going to say. Try to understand their point of view, not just wait for your turn to talk. Ask clarifying questions like, "So, if I understand correctly, you're saying...?"

- Debate (Respectfully!): Engage in respectful debates with friends or family. Try to argue for a point of view that you don't necessarily agree with. This helps you see things from a different angle and strengthens your ability to construct arguments from multiple viewpoints.

- Read Diverse Voices: Read books, articles, or blogs written by people from different backgrounds, cultures, and with different viewpoints than your own. This exposes you to new ideas and helps you understand different ways of thinking.

- People Watch (Ethically!): Observe people in your school or community. Try to imagine what their lives are like and why they might be acting a certain way. This isn't about judging them; it's about developing empathy and understanding different perspectives.

- Role-Play: Try role-playing different scenarios with friends. Take on a character with a different perspective and try to act and think like that person.

- "Yes, and..." Improv: Try playing improv games like "Yes, and..." This forces you to build on other people's ideas and see where they lead, even if they're unexpected.

- Reflect on Your Own Views: Take time to think about why you believe what you believe. Where did your opinions come from? Are there other perspectives you haven't considered? Journaling can be helpful for this.

By practicing these things regularly, you'll start to naturally consider multiple perspectives in your daily life, making you a better problem-solver, communicator, and a more understanding person overall. It's like leveling up your social skills!

ASTS. Different Scales of Abstraction

- Think "Big Picture" and "Small Details": When you're working on a project, whether it's for school or something you're doing for fun, try to think about both the overall goal and the individual steps needed to get there. Don't get lost in the details so much that you forget what you're trying to achieve, and don't just focus on the big picture without considering how you'll actually make it happen.

- Break Things Down: Take something complex, like planning a party or learning a new skill, and break it down into smaller, more manageable parts. This helps you see how the different pieces fit together and makes the whole thing less overwhelming. It's like leveling up in a game – you don't try to beat the final boss right away, you break it down into smaller quests.

- "Zoom In, Zoom Out": Practice shifting your perspective. Think about your school – you can look at it as a whole (lots of students, teachers, classrooms), or you can zoom in on your class, your group of friends, or even just your own experience. Try to understand how these different levels connect.

- Map it Out: Creating mind maps or diagrams can be helpful for visualizing different scales of abstraction. Start with the big idea in the center and then branch out to the smaller details.

- Analyze Systems Around You: Start noticing systems in your everyday life, like the public transportation system, the school's grading system, or even how your family organizes chores. Try to understand how these systems work at different levels, from the individual parts to the overall function.

By consciously practicing these techniques, you'll become more comfortable thinking at different scales of abstraction, which is a crucial skill for understanding complex problems and developing effective solutions. You'll start to see how everything is connected and how the details contribute to the bigger picture.

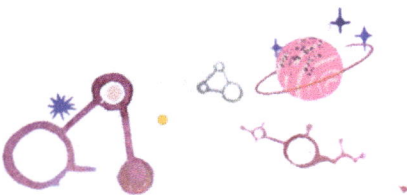

ASTS. Interconnections, Intrarelationships, and Dependencies

- Analyze Team Dynamics: Pay attention to how teams work, whether it's in sports, group projects, or even just how your family gets things done. Who depends on whom? What happens if someone doesn't do their part? Try to understand the relationships between the different members.

- Map Out Connections: When you're trying to understand something complex, like a current event or a scientific concept, try drawing a diagram or a mind map to show how the different parts are connected. Visualizing these connections can make them easier to understand.

- Play Strategy Games: Video games that require teamwork and strategy are great for developing this skill. Think about how your actions affect your teammates and how you need to coordinate to achieve a common goal.

- Consider Cause and Effect: Practice thinking about cause and effect in your daily life. If you do this, what might happen? If someone else does that, how might it affect others? This helps you understand the dependencies between different actions and outcomes.

- Look for Patterns: Try to identify patterns in how things work. Do certain actions always lead to certain results? Recognizing these patterns helps you understand the underlying connections and dependencies.

- Systems Thinking in Conversations: When talking with friends or family, try to think about the bigger context of what you're discussing. How are the different parts of the conversation related? Are there unspoken dependencies influencing the discussion?

By paying attention to these connections and dependencies in your daily life, you'll start to develop a better understanding of how systems work and how you can influence them.

ASTS. Dynamic Behavior

- Observe Changes Over Time: Pay attention to how things change over time. Track the weather, observe how a plant grows, or even notice how your own mood changes throughout the day. This helps you understand that change is a constant.

- Experiment and Observe: When you're trying something new, like learning a skill or trying a recipe, don't be afraid to experiment and see what happens. Observe how the system changes in response to your actions.

- Play Simulation Games: Games that simulate real-world systems, like city-building games or economic simulations, can help you understand how different factors interact and how systems change over time.

- Consider Different Scenarios: Practice thinking about "what if" scenarios. What if this happens? What if that changes? How might the system respond? This helps you anticipate dynamic behavior.

- Analyze Trends: Look for trends in the data around you. Are certain things increasing or decreasing? Are there patterns in how things change? This can help you understand the forces that are driving dynamic behavior.

- Reflect on Your Own Behavior: Think about how your own behavior changes in different situations. How do your emotions affect your actions? How do other people's actions affect you? Understanding your own dynamic behavior can help you better understand the behavior of other systems.

By actively observing and analyzing the dynamic behavior of systems around you, you'll become better at anticipating change, adapting to new situations, and making informed decisions. You'll start to see how everything is connected and how even small changes can have big consequences.

ASTS. Stock, Flow, and Delay

- Track Your Spending: Keep track of how much money you have (stock), how much you spend (flow), and how long it takes to save up for something you want (delay). This helps you understand how your financial resources change over time.

- Observe Resource Use: Pay attention to how resources are used in your daily life. How much water do you use when you shower? How much time do you spend on social media? This helps you understand the flow of resources.

- Plan Ahead: When you're planning something, like studying for a test or completing a project, think about the resources you'll need (time, materials, etc.) and how long it will take to acquire them. Consider the delays involved.

- Think About Feedback Loops (Related to Feedback ASTS): If you change the flow of something, how will it impact the stock? If you spend more money, your savings will decrease. If you study more, your knowledge will increase. Understanding these feedback loops is key to managing stocks, flows, and delays.

- Consider Environmental Impact: Think about how resources are used in your community and the impact of delays. How long does it take for trash to decompose? How long does it take for pollution to affect the environment?

- Analyze Game Mechanics: Many video games incorporate stock, flow, and delay mechanics. Resource management games, simulations, and even some strategy games require you to think about how resources accumulate, deplete, and how delays affect your progress.

By paying attention to stocks, flows, and delays in your everyday life, you'll develop a better understanding of how systems behave over time and how you can manage resources effectively.

ASTS. Feedback

- Reflect on Your Actions: After you do something, take a moment to think about the outcome. What happened? Was it what you expected? What did you learn from the experience? This is a form of self-feedback.

- Seek Feedback from Others: Ask for constructive criticism from friends, family, teachers, or coaches. How could you improve? What did you do well? Be open to hearing both positive and negative feedback.

- Pay Attention to Your Emotions: Your emotions can be a form of feedback. If you feel stressed or overwhelmed, it might be a sign that you need to adjust your approach to something.

- Experiment and Observe: When you try something new, pay close attention to the results. What worked? What didn't? Use this feedback to refine your approach.

- Analyze Game Dynamics: Video games are full of feedback loops. Pay attention to how your actions affect the game world and use that feedback to improve your gameplay.

- Use Feedback to Improve: Don't just collect feedback; use it to make changes. If you receive constructive criticism, try to implement it. If something didn't work, try a different approach.

By actively seeking and using feedback in your daily life, you'll become more adaptable, learn from your mistakes, and continuously improve your skills. You'll start to see how feedback loops influence everything around you.

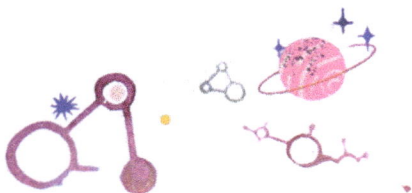

ASTS. Non-linear Relationships

- Observe Real-World Examples: Look for examples of non-linear relationships in your life. Think about how exercise affects your fitness, how social media use affects your mood, or how studying affects your grades. Are the changes always proportional?

- Experiment and Analyze: When you're trying something new, pay attention to the results. Did the outcome match your expectations? If not, what might have caused the unexpected result?

- Play Strategy Games: Many video games involve non-linear relationships. Changing one variable, like the amount of resources you invest in a particular unit, can have unpredictable consequences for your overall strategy.

- Consider Complex Systems: Think about complex systems like the weather or the economy. Small changes in one area can have ripple effects throughout the system, leading to unexpected outcomes.

- Read About Chaos Theory: For a more advanced understanding, you can explore the concept of chaos theory, which studies how small changes can lead to large-scale unpredictability in complex systems (the "butterfly effect").

Developing an understanding of non-linear relationships will help you make better predictions, anticipate unexpected consequences, and make more informed decisions in a complex world.

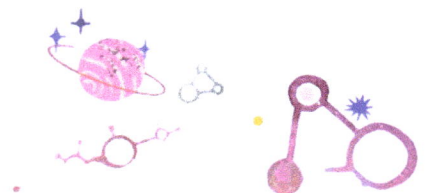

ASTS. Mental and Formal Models

- Explore Simulations and Coding: Use online simulations, games or basic coding to experiment and directly test the assumptions embedded in your mind. This allows you to observe how changes in variables affect systems and can help you refine your mental models, revealing potential flaws or insights.

- Compare Formal Models: Explore the cross-applicability of formal models by showcasing how similar underlying mathematical or algorithmic principles, like exponential growth or network analysis, appear in diverse fields like finance, climate science, and social media. This encourages you to identify shared structures and the transfer of concepts between seemingly disparate domains.

- Self-Questioning: Actively challenge your assumptions and beliefs as you would explore different subjects. Systematically scrutinize the evidence supporting your thoughts and deliberately seek alternative viewpoints, fostering a more objective and nuanced understanding of your own internal landscape.

- Intentional, Varied Reading: Actively seek out subjects beyond your comfort zone, like history, art, or philosophy. Engage deeply by taking notes and connecting concepts, aiming for consistent exposure through set reading goals.

- Cultural Immersion: Explore diverse cultures through travel, friends, documentaries, and online communities, practicing active listening and seeking out varied news sources. Language learning and attending cultural events can also broaden your worldview.

- Experiential Diversification: Step outside your routine by trying new hobbies, engaging in creative activities, and volunteering. Reflect on these experiences to integrate them into your understanding of the world.

These strategies empower you to uncover hidden assumptions, understand your own mental models, and effectively apply formal models to navigate real-world situations, bridging the gap between theory and practice and enabling you to navigate challenges with greater clarity and effectiveness.

ASTS. System Structure and Boundary

- Analyze Organizations: Think about different groups you're a part of – your family, your friend group, a sports team, a club. What's the structure of each group? Who's in charge? How are decisions made? Where are the boundaries – who's considered a "member" and who's not?

- "Draw" the System: When you're trying to understand something complex, try drawing a diagram that represents the system's structure. Use boxes for the different parts and lines to show how they're connected. Define the boundaries – what's included in your diagram and what's not.

- Consider Context: Think about how the context affects a system. Your school exists within a larger community, and that community influences the school. Understanding the boundary between the school and the community is important.

- "What's In, What's Out?": Practice defining the boundaries of different systems. Think about a conversation you had with a friend. What was "in" the conversation (the topics you discussed) and what was "out" (things you didn't talk about)?

- Social Media Ecosystem: Think about your personal social media ecosystem, including platforms, connections, and information flows. Then, discuss how changes to algorithms, privacy settings, or your own usage habits impact your online experience and social interactions. This activity explores digital system boundaries and the influence of external factors on personal networks.

By practicing identifying the structure and boundaries of systems in your daily life, you'll become better at understanding how things are organized and how different systems relate to each other.

ASTS. Conceptual Modeling

- Explain Things Simply: Practice explaining complex ideas or systems to others in a simple way. Try to capture the main points without getting bogged down in the details. This helps you develop your ability to create effective conceptual models.

- Use Analogies and Metaphors: Relating complex concepts to familiar things can make them easier to understand. For example, you could compare the internet to a network of roads.

- Create Mind Maps: Mind maps are a great way to visually represent the key concepts and relationships within a system. Start with the main idea in the center and then branch out to related concepts.

- Tell Stories: Stories can be a powerful way to communicate complex information. Try telling a story that illustrates the key elements of a system.

- "Elevator Pitch": Imagine you only have a few seconds to explain a complex idea to someone. What are the most important things you would say? This helps you distill information down to its essence.

- Use Visual Aids: Drawings, diagrams, and other visual aids can be helpful for creating conceptual models. Even simple sketches can help you clarify your thinking.

By practicing creating conceptual models in your daily life, you'll become better at simplifying complex information, communicating effectively, and understanding the core principles of different systems.

ASTS. Prospection and Prediction

- "What If?" Scenarios: Practice thinking about "what if" scenarios. What if I study for an extra hour? What if I try a different approach? How might these actions affect the outcome?

- Plan Ahead: When you're planning something, like a party or a project, think about the potential challenges and how you might overcome them. This is a form of prospection and prediction.

- Analyze Past Events: Look back at past events and try to understand why they happened. What factors contributed to the outcome? This can help you make better predictions in the future.

- Play Strategy Games: Games that require planning and strategy are great for developing prospection and prediction skills. Think about chess, strategy board games, or even some video games.

- Follow the News: Staying informed about current events can help you understand the forces that are shaping the future. Try to predict how different events might unfold.

By practicing prospection and prediction in your daily life, you'll become better at anticipating potential outcomes, making informed decisions, and preparing for the future.

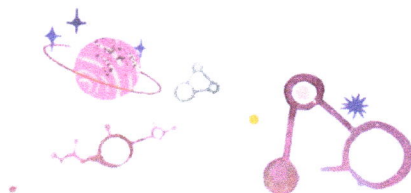

ASTS. Hypothetical and Inferential Consideration

- Solve Puzzles and Riddles: Puzzles and riddles are great for developing inferential thinking. You need to use the information you have to draw logical conclusions.

- Play "What If?" Games: Make a habit of playing "what if" games with yourself or with friends. What if I took a different route to school? What if I tried a different sport? This helps you explore different possibilities.

- Analyze Arguments: When someone is making an argument, try to identify the underlying assumptions and the evidence they're using to support their claims. Are their conclusions logical?

- Read Critically: When you're reading, don't just passively absorb the information. Question what you're reading, look for hidden assumptions, and try to draw your own conclusions.

- Consider Different Interpretations: When you're faced with a situation, try to think about different ways it could be interpreted. What are the possible explanations? What evidence supports each interpretation?

By practicing hypothetical and inferential thinking in your daily life, you'll become better at analyzing information, drawing logical conclusions, and exploring different possibilities. You'll be able to see beyond the surface and understand the underlying dynamics of complex situations.

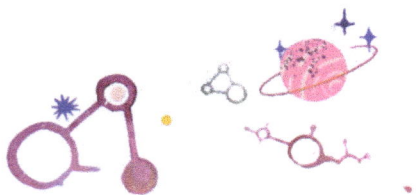

ASTS. Paradoxical and Ambiguity Tolerance

- Embrace "Gray Areas": When faced with a decision, resist the urge to see it as a simple "yes" or "no" choice. Consider the different options and the potential consequences, even if they're not clear-cut.

- Challenge Assumptions: Question your own assumptions and biases. Are you seeing the situation clearly, or are you making assumptions based on limited information?

- "Both/And" Thinking: Practice thinking in terms of "both/and" rather than "either/or." For example, you can be both creative and analytical, or you can care about your friends and also prioritize your studies.

- Learn to Tolerate Discomfort: Uncertainty can be uncomfortable, but it's a part of life. Practice being okay with not knowing all the answers and making decisions based on the best information you have available.

- Reflect on Past Experiences: Think about times when you faced ambiguity or paradoxes. How did you handle those situations? What did you learn from them?

Developing tolerance for paradox and ambiguity will help you navigate the complexities of life with greater confidence and resilience. You'll be able to make better decisions in uncertain situations and adapt more easily to change.

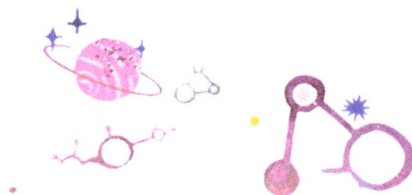

ASTS. Creativity

- Brainstorming: Practice brainstorming ideas, even if they seem silly or impractical at first. The goal is to generate a large quantity of ideas, and then you can refine them later.

- Try New Things: Step outside your comfort zone and try new activities, hobbies, or even foods. New experiences can spark new ideas.

- Observe the World Around You: Pay attention to the details of the world around you. Look for patterns, connections, and inspiration in unexpected places.

- Doodle and Sketch: Even if you don't consider yourself an artist, doodling and sketching can help you visualize ideas and explore different possibilities.

- Don't Be Afraid to Fail: Creativity involves experimentation and risk-taking. Not every idea will be a winner, and that's okay. Learn from your mistakes and keep trying.

- Collaborate and Share Ideas: Bouncing ideas off other people can lead to new insights and perspectives. Work with others to develop your creative ideas.

By nurturing your creativity in these ways, you'll become more innovative, resourceful, and better equipped to solve problems in unique and effective ways. You'll discover your own creative potential and be able to express yourself in new and exciting ways.

Appendix D. Brain Teaser Solutions

The Mysterious Message: You find a note with the following symbols: △□○◇. Each symbol represents a letter. △ = A. □ = T. ○ = O. What word does ◇ represent? *(Answer: It's impossible to know without more information but was intended to encourage you to ask clarifying questions and recognize the limits of given data.)*

The Lost Kitten: A kitten is stuck at the top of a tall tree. Three people offer to help. A firefighter suggests a ladder. A construction worker suggests a crane. A child suggests something simpler. What does the child suggest? *(Answer: Calling the kitten's mother. This encourages thinking outside the box and considering different perspectives.)*

The Empty Box: You have a box. It's empty. Can you make it full without putting anything inside it? *(Answer: Yes, fill it with air/light. This encourages creative thinking and challenging assumptions about "full.")*

The Repeating Riddle: What is always coming, but never arrives? *(Answer: Tomorrow. This classic riddle encourages abstract thinking and understanding of time.)*

The Mismatched Socks: You have a drawer full of socks. You have 10 white socks and 10 blue socks. It's dark, and you can't see the colors. How many socks do you need to take out to guarantee you have a matching pair? *(Answer: Three. This encourages logical deduction and understanding of probability.)*

The Broken Bike: Your friend's bike chain broke while you're miles from home. You have no tools. How do you get the bike moving again (even if it's not perfect)? *(Answer: Look for something that can act as a temporary chain link, like a sturdy stick, a piece of wire, or even a strong vine. This encourages resourcefulness and thinking about the interconnected parts of a system.)*

The Leaky Roof: A storm is coming, and there's a small leak in your roof. You don't have time for major repairs. What's a quick, temporary fix using common household items? *(Answer: Use a plastic sheet or garbage bag and secure it over the damaged area with tape, bricks, or heavy objects. This focuses on practical problem-solving with available resources and understanding the system of weather and shelter.)*

The Confusing Code: You find a coded message: 1-1-2-3-5-8-13-21. What does it mean? *(Answer: This is the Fibonacci sequence. Each number is the sum of the two preceding ones. This tests pattern recognition and understanding of mathematical systems.)*

The Island Escape: You're stranded on a deserted island with a box of matches. You need to build a signal fire, but all the wood you find is damp. How do you start the fire? *(Answer: Look for dry kindling, such as birch bark, dried grass, or pine needles. You could also try splitting larger pieces of wood to expose drier inner surfaces. This emphasizes resourcefulness and adapting to environmental systems.)*

The Growing Problem: A patch of weeds in your yard doubles in size every week. If it takes 8 weeks to cover the entire yard, how long will it take to cover half the yard? *(Answer: 7 weeks. This tests understanding of exponential growth and working backward in a system. Many people incorrectly say 4 weeks.)*

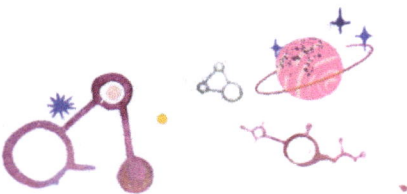